# Carbon Monoxide Sensing Technologies

**Gurleen Kaur Gulati, Loveleen Kaur Gulati, and Satish Kumar***

Department of Chemistry, St. Stephen's College, University Enclave,
Delhi 110007, India

*satish@ststephens.edu

Edited by

**Inamuddin**

Department of Applied Chemistry, Zakir Husain College of Engineering and Technology,
Faculty of Engineering and Technology, Aligarh Muslim University, Aligarh-202 002,
India

Published by **Materials Research Forum LLC**
Millersville, PA 17551, USA

Published as part of the book series
**Materials Research Foundations**
Volume 94 (2021)
ISSN 2471-8890 (Print)
ISSN 2471-8904 (Online)

Print ISBN 978-1-64490-120-5
ePDF ISBN 978-1-64490-121-2

This book contains information obtained from authentic and highly regarded sources. Reasonable efforts have been made to publish reliable data and information, but the author and publisher cannot assume responsibility for the validity of all materials or the consequences of their use. The authors and publishers have attempted to trace the copyright holders of all material reproduced in this publication and apologize to copyright holders if permission to publish in this form has not been obtained. If any copyright material has not been acknowledged please write and let us know so we may rectify in any future reprint.

Distributed worldwide by

**Materials Research Forum LLC**
105 Springdale Lane
Millersville, PA 17551
USA
http://www.mrforum.com

Printed in the United States of America
10 9 8 7 6 5 4 3 2 1

# Table of Contents

The release of poisonous gases like carbon monoxide (CO) is an extremely serious problem that concerns the global community. In order to prevent the damage to the human population, it is essential to develop easier methods for the detection of carbon monoxide concentration in the environment. During the last several decades, various methods have been developed to monitor the presence of carbon monoxide. The chapter discusses the techniques developed for sensing carbon monoxide. Various techniques have been compared and discussed, which can provide the reader an overview of the techniques best suited for the recognition and monitoring of carbon monoxide polluting our environment.

## 1. Introduction

The increase in the global population is driving a number of industrial activities that release a variety of toxic gases into our environment. The release of toxic gases affects the health of the human population. Among various toxic gases released through industrial and domestic activities, carbon monoxide represents a clear danger to humans and the ecology. Carbon monoxide (CO) is a highly toxic atmospheric gas pollutant, which is a stable oxide of oxygen. It is colorless, odourless, non-flammable, non-corrosive and non-irritating in nature [1]. Due to these physicochemical properties, it is difficult to be detected in the ambient atmosphere until it is too late. Hence, carbon monoxide is also termed as a *"silent killer"*. It has also been named as *"the unnoticed poison of 21st century"* [2-4].

It is generated in the environment by incomplete combustion of carbon-containing compounds [1, 3, 5, 6]. It is also produced endogenously in the body and is recognized as a gasotransmitter in the central nervous system (CNS). It serves as a physiologically significant signalling molecule at non-toxic concentrations [7-10]. The external and endogenous sources of carbon monoxide emission are listed in Table 1 [11-14].

The atmospheric concentration of CO is less than 0.001%. As a result of atmospheric and endogenous CO, the blood carboxyhemoglobin (COHb) levels range from 1-3% in non-smokers & 10-15% in smokers [15]. Carboxyhemoglobin saturation level can reach 8-10% after 8h exposure to 50 ppm CO [16].

*(i) Mechanism of CO toxicity*: CO is taken up by lungs (Alveoli) and then transferred into the blood stream by pulmonary absorption. Therefore, carbon monoxide reversibly binds with hemoglobin (Hb) to form COHb [17]. The quantity of gas absorbed in the body depends on the concentration of CO in the environment as well as the exposure duration [18]. The affinity of Hb for CO is almost 210-250 times higher than that for oxygen. Therefore, inhalation of CO leads to increased levels of carboxyhemoglobin

(COHb) in the blood thereby reducing both, the oxygen-carrying capacity and oxygen releasing capacity of theme [14, 15]. This generates a medical condition called tissue hypoxia as well as cellular toxicity.

Apart from binding to hemoglobin, carbon monoxide is also known to interact with myoglobin and cytochrome $a_3$ in mitochondria [18].

***Table 1****. Sources of carbon monoxide [11-14].*

|   | **ENDOGENOUS SOURCES** |
|---|---|
| 1. | Normal heme catabolism by heme oxygenase |
|   | **EXTERNAL SOURCES** |
| 1. | Automobile Exhaust |
| 2. | Combustion Appliances involving partial oxidation of carbonaceous fuel (oil, coal, wood, charcoal, kerosene, etc.):<br>→Propane-powered engines (forklifts, chain saws)<br>→Catalytic radiant heaters<br>→Portable generators<br>→Kerosene heaters<br>→Natural gas appliances<br>→Hibachi cookers |
| 3. | Gas log fireplaces |
| 4. | Fires |
| 5. | Paint strippers |
| 6. | Spray paint |
| 7. | Cigarette, cigar & pipe smoke |
| 8. | Boat exhaust |
| 9. | Indoor grills |
| 10. | Camp stoves |

***(ii) Health complications due to CO poisoning***: The organs of the body most affected by carbon monoxide poisoning are the ones with high metabolic activity like the heart and areas of the brain [18, 19]. Neurological effects of mild CO poisoning are headache, dizziness, lethargy, depression, syncope, seizures, impaired vision, convulsions, coma and many more [16, 18]. These effects worsen if CO exposure is chronic, then it can lead to blindness, deafness, pyramidal signs, unconsciousness, and death. Neuropsychiatric effects include euphoria, confusion, impaired judgment, etc. Cardiovascular disorders resulting from CO toxic damage involve conditions like life-threatening dysrhythmia, myocardial infarction and cardiac arrest [6, 18-20]. Acute & chronic exposure can lead to parkinsonism and delayed chorea in some cases [21]. The severity of such hazardous health consequences of CO exposure lead to vast research in the late 20th century in the field of sensors to identify this otherwise undetectable gas.

## 2.    Types of gas sensing devices

During the last several decades, different types of gas sensing techniques have been developed using various methodologies and materials. The sensors that had been fabricated and analyzed for carbon monoxide sensing can be broadly categorized into:

  (i) Semiconducting metal oxide-based sensors
  (ii) Carbon nanotubes semiconductors
  (iii) Conducting polymeric thin films
  (iv) Sensors based on colorimetric detection
  (v) Non-dispersive infrared sensors
  (vi) Electrochemical sensors:
        (a) Potentiometric sensors
        (b) Amperometric sensors
  (vii) Photoacoustic detection-based sensors

The following section will provide a discussion of the family of sensors listed above in some detail.

### (i) Semiconducting metal oxide-based carbon monoxide gas sensors

Sensing based on semiconductor metal oxides involves recognizing changes in electrical conductivity response of the sensor against the gaseous stimuli [22-29]. The metal oxide semiconductor type gas sensors include a wide variety of oxides being used in these devices like $SnO_2$, $TiO_2$, $ZnO$, $In_2O_3$, $CeO_2$, $MoO_3$, etc. Several modifications have been incorporated by doping different elements (Al, Ca, Au, Pt, Pd) in the semiconductor to enhance its electrical conductivity and response sensitivity [22, 30-35]. On the grounds of the type of fabrication technique, these devices can be of sintered bulk, thin film or thick film type [36, 37]. There has been an increasing practice of modelling these sensors into distinct nano-designs like nanorods, nanosheets, nanotubes, nanowires among many more. The transformation to nano-scale greatly enhances the surface/volume ratio parameter which results in increased rates of interaction between gas analyte molecules and the sensor device. This is how nano-fabrication helps to improve the sensitivity and selectivity of the sensor response [38, 39].

The mechanism of CO sensing by this type of metal oxide sensors was proposed by Sheikhet et al. [40]. The authors have investigated the $TiO_2$ (anatase) semiconductors for their response towards carbon monoxide gas. An anatase is a metastable form of titanium oxide. The investigation indicated that the electrical response process can be explained in two steps. In the first step, the adsorption of oxygen (from the atmosphere) on to the grain surface and its conversion to negatively charged species by extraction of electrons from the matter. In the second step, the interaction of CO molecules (analyte) with the surface adsorbed oxygen species to form $CO_2$, in turn, giving the electrons back to the conduction band resulting in increased conductivity response of the sensor.

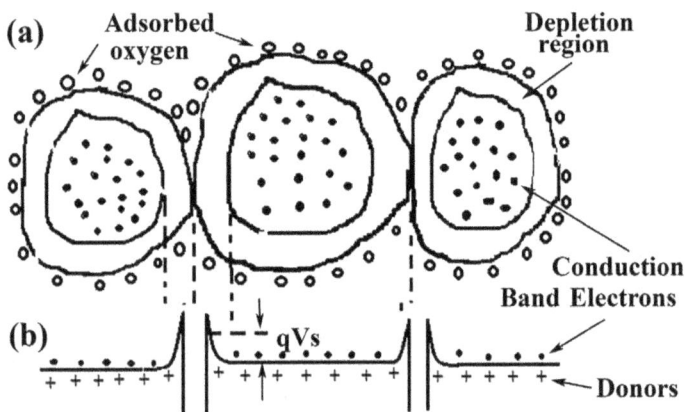

*Fig. 1.(a) Depletion layer, (b) Energy Barrier in qV created between the grains on the sensor surface [40].*

In the intermediate step, the generation of depletion region (depicted in Fig. 1) [40] around the grains occurs due to the continuous excavation of electrons from the surface, which acted as a barrier for the intergranular movement of electrons. The inter-activity of the surface with the carbon monoxide gas molecules helped to decrease this barrier height and increase the conductivity. Hence the barrier magnitude can be directly linked to and exploited for altering the sensitivity of the sensor response.

Appreciable modifications in $TiO_2$ sensors have been witnessed in literature like yttria doping [41], lanthanum oxide – copper oxide doping [42], and Nb-doped $TiO_2$ nano-structure [43]. Still, $TiO_2$ based carbon monoxide sensors are fewer in count in comparison to the tin oxides-based ones, which are discussed at length in this chapter.

## (a) Tin(IV) oxide (SnO₂)based sensors

Tin(IV) oxide is an *n*-type semiconductor, which has been extensively used to design toxic gas sensors of a wide variety, CO detectors are some among them. The reason for the comparatively more usage is their easy construction [44], large bandgap (3.6 eV at 300K) [45-47], higher stability [48] and cost-effectiveness [49, 50].

This type of gas sensor generally operates at high temperatures to achieve higher stability, high sensitivity and quick response times [51, 52]. The annealing is also done to manage intrinsic defects and grain size of the sensor surface [52]. Several efforts have been made to lower this operating temperature by the inclusion of catalytic metal oxides ($MoO_3$, $CuO$, $Fe_2O_3$) [51, 53-55] so that their power consumption can be minimized. Various other combinations of dopants, noble metals and cations have been incorporated to obtain better sensing characteristics.

Harrison et al. [56] have illustrated the working principle of tin dioxide sensors. The investigation was aimed to study the direct impact of the surface adsorption process on the electrical signal of the bulk oxide. The transmission infrared spectroscopy was used to closely understand the surface reaction between $SnO_2$ and adsorbed (oxygen, carbon monoxide) species. The chemical entities observed during the experiment are diagrammatically represented in Fig. 2 [56].

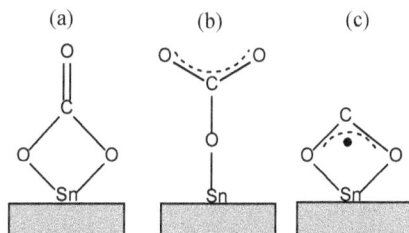

*Fig. 2. Formation of surface species on tin dioxide in CO/ air environment. (a) Bidentate carbonate (characteristic infrared bands; 1,585, 1,223 cm⁻¹), (b) Unidentate carbonate (1,430, 1,370 cm⁻¹), (c) Carboxylate (1,540, 1,300 cm⁻¹) [56].*

It has been particularly emphasized that the sensing temperatures greatly affect the adsorption intensities of species and thus also influence the electrical conductance response generated on CO sensing. The adsorption peaks of the surface moieties are showing maximum abundance at around 570 K (Fig. 3) [56]. It has been concluded that high pre-treatment temperature (approx. 570 K) caused dihydroxylation of the sensing surface facilitating the easy adsorption of oxygen moieties (from the atmosphere) and their reaction with CO molecules.

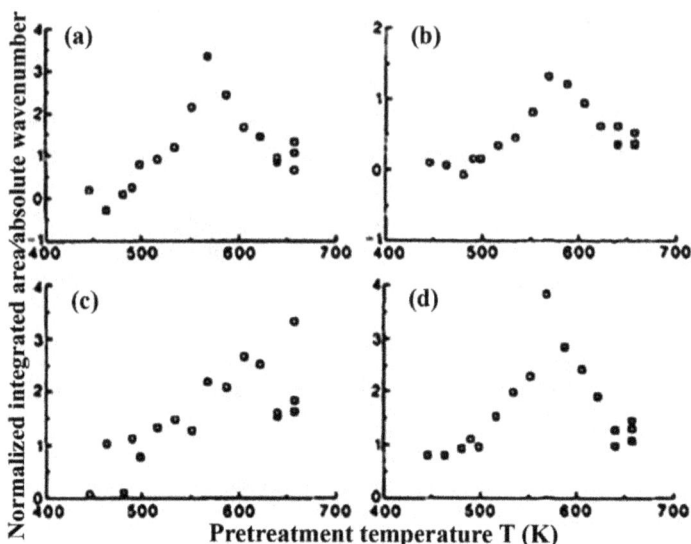

***Fig. 3****. Graphs plotted between the integrated peak areas and sensor pre-treatment temperature for the bands observed at (a) 1,585 cm⁻¹, (b) 1,223 cm⁻¹ (c) 1,300 cm⁻¹, (d) 1,430 cm⁻¹.Reproduced here with the permission of Springer Nature [56].*

At this temperature, desorption of $CO_2$ was favoured which promoted the replenishment of electrons into the conduction band of metal oxide resulting in a hike in the electrical signal of the sensor. It has been demonstrated that the $SnO_2$ sensor sensitivity enhances at higher pre-treated temperatures [56].

Tsai et al. [36] have described the modeling of thick film $SnO_x$ sensor (Fig. 4) [36] comprising the calcium oxide and niobium oxide as dopants along with Pt as a catalyst [36]. The doped metal oxide formed the sensing sheet, which was prepared by a film deposition technique popularly known as metallo-organic decomposition (MOD). Screen-printed electrodes and heating elements were utilized.

***Fig. 4.*** *Sensors prepared by a thick film deposition method. Reproduced here with the permission of Elsevier [36].*

The sensing experiments were performed at different sensor temperatures by varying the power of the heating element to establish a relation between the two that higher temperatures required more watts of power. The developed sensor has been demonstrated to detect a minimum of 30 ppm concentration of CO. The sensitivity of the CO sensor was good between 30 to 2800 ppm with almost linear variation up to 400 ppm (Fig. 5) [36].

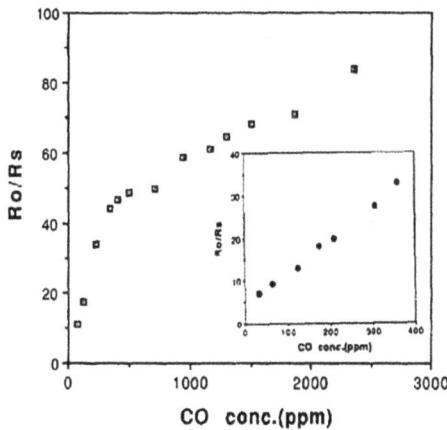

***Fig. 5.*** *The plot of CO sensitivity of sensor vs concentration of CO (ppm) in a range of 30-2800 ppm. The inside plot shows approximate linearity in the curve below 400 ppm. Reproduced here with the permission of Elsevier [36].*

The problem encountered due to the interference from hydrogen and ethanol at 150°C was solved by lowering the sensing temperature to 50°C along with a change in catalyst from Pt to Pd.

Another innovation towards developing a selective & sensitive CO sensor was made by Sharma et al. [52] who devised a Cu-doped $SnO_2$ thin film followed by deposition of an even thinner Pt film over it to give structure like $SnO_2$–Cu/Pt (Fig. 6) [52]. The films were formulated by a method known as Radio Frequency Sputtering and were characterized using scanning electron microscope (SEM) and X-ray photoemission spectroscopy (XPS).

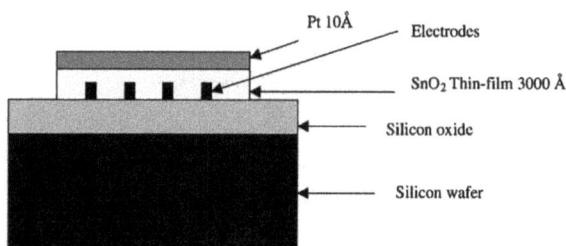

**Fig. 6**. *Design of the sensor. Reproduced here with the permission of Elsevier [52].*

In the study, platinum has been used to increase the resistance of thin-film as it allowed more adsorption of atmospheric oxygen during the annealing process, resulting in greater activation & speedy response of the sensor towards analyte (carbon monoxide) gas molecules [52, 57]. Pt also served to improve the catalytic activity for CO sensing at lower sensor temperature. The sensor showed a better response at higher CO concentrations. The response behavior has been observed to increase with an increase in the operational temperature of the thin film. At 270-320°C, a response time of 5-7s has been observed.

Cerium oxide has emerged as a promising alternative to the existing additives being used in metal oxide sensors. It displays greater sensitivity for CO [58] & possesses the ability to dissociate carbon monoxide molecules on metal/ceria surface [59, 60]. In addition, $CeO_2$ is also considered to be a highly porous material ideal to be used in sensing applications [61-63]. Khodadadi et al. [59] have used 10% $CeO_2$ in $SnO_2$ semiconducting sensor to obtain better sensitivity for CO along with sufficient selectivity.

Further probing in this domain led to the discovery of highly sensitive CO sensors by Durrani et al. [61]. The physical vapor deposition approach has been utilized to prepare mixed thin films of $SnO_2$ and $CeO_2$ on different substrates (Ta, fused silica, alumina) for different types of analysis. For gas sensing, films grown on alumina substrate were analyzed, its thickness was 220 nm. The $SnO_2$-$CeO_2$ mixed sensor has been found to be responsive to carbon monoxide concentrations as low as 5.0 ppm (Fig. 7) [61] at an operating temperature of 430°C & 1.5V bias voltage. The response time (26 s) and the recovery time (30 s) of the sensor were also found to be good enough for rapid sensing.

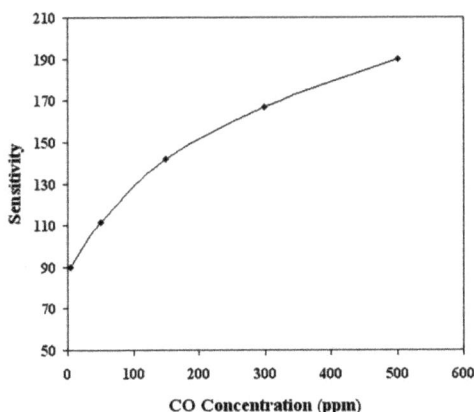

***Fig.** 7. The SnO2-CeO2 sensor response to various CO concentrations at 430 ⁰C. Reproduced from reference [61].*

All the $SnO_2$ sensors discussed above worked at very high operating temperatures, which were achieved using some heating element along with the sensor. With the use of heating devices, high power utilization was a problem. So, A. Salehi [44] has tried to solve this complication by creating a self-heated $SnO_2$ sensing device for carbon monoxide gas. The sensing film inherently served as a heater (ac voltage could be directly attached to the film), simplifying the sensor design as well as reducing the power dissipation. The $SnO_2$ film was developed on a Pyrex substrate using chemical vapor deposition (CVD). The sensitivity of this film witnessed many fold increase as compared to the un-doped $SnO_2$ traditional sensors, the reason being the extensive formation of grain dispersions (leading to enhanced porosity of the film) upon application of ac voltage directly to the film. Due to this, facilitated penetration of CO gas molecules occurred uniformly into the whole film deep inside, as against a surface phenomenon in conventionally heated sensors. The additional benefits of this invention were reported to be low expense, easy design and compact size ideal for microelectronic applications [44].

The type of film deposition technique also plays a vital role in determining the performance of the sensor film towards a particular gas analyte. Spray pyrolysis is one such technique that has repeatedly been employed to develop thick/thin films, porous films, multi-layered films as well as powders [64]. It exhibits a simplified & cost-efficient method to prepare films of a wide range of compositions without the need for high-grade chemicals. In this method, film deposition is done when a solution is sprayed onto a heated surface where the reactants combine to form a chemical compound [65]. The constituents are chosen such that the by-products formed during the reaction are volatile at deposition temperature, leaving behind the film of the desired compound.

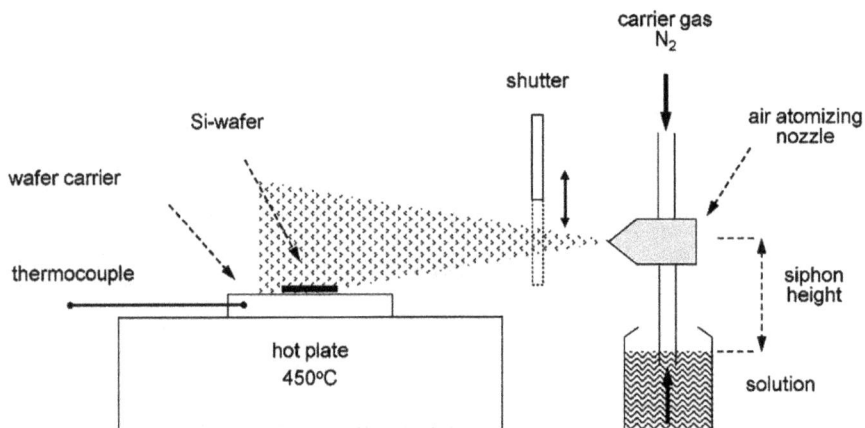

**Fig. 8.** *Spray pyrolysis scheme of film deposition. Reproduced here with the permission of Elsevier [66].*

Tischner et al. [66] have utilized this technique for devising an ultrathin $SnO_2$ film sensor that evolved on oxidized Si substrate. The process is schematically described in Fig. 8 [66]. Scanning electron microscopy (SEM), atomic force microscopy (AFM) and X-ray photoelectron spectroscopy (XPS) was employed to closely examine the structural and chemical composition of the film. The AFM results are presented in Fig. 9 [66].

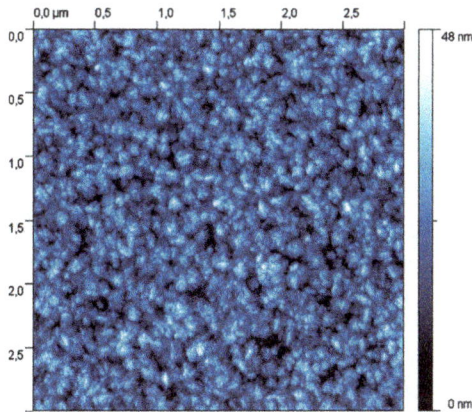

***Fig. 9****. The 80 nm thick layer of SnO₂ observed using AFM. Reproduced here with the permission of Elsevier [66].*

The sensor has been found to be highly responsive to low CO concentrations (5 to 20 ppm). However, the sensitivity towards carbon monoxide has been observed to decrease in the high concentration range (up to 200 ppm) (Fig. 10) [66].

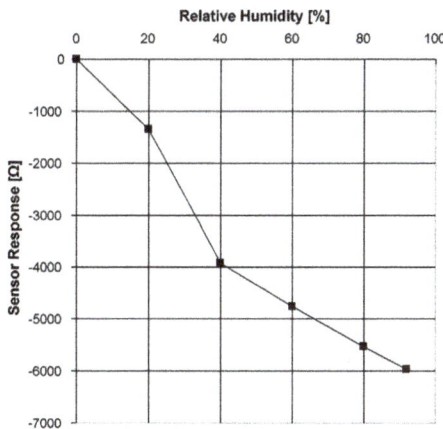

***Fig. 10****. A plot of sensor response against the concentration of CO gas at 250 °C. Reproduced here with the permission of Elsevier [66].*

The only shortcoming observed in the device has been the interference, caused due to the presence of humidity, during the selective detection of carbon monoxide gas as both (moisture and CO) caused a decrease in the sensor response thereby producing signals at the detector.

Doping is yet another modification in the composition of metal oxide sensors in order to achieve improved sensing attributes. The dopants can be single transition metal species or a combination of two. The metal additives act as catalysts by assisting in the interaction of gas molecules and the sensor surface [51]. These are also known to lower the temperature of operation of the sensor and increase its response sensitivity [48, 67]. Metal dopants are considered favorable in refining the electrical conduction properties of the sensor as these furnish varying electronic structures, which aid in increasing surface states, charge carriers & active sites for better sensing [51, 68, 69].

Wang et al. [51] have reported the construction of vanadium assisted tin oxide nanostructure for CO recognition at low temperature. The composition, surface state, and structure of the metal oxide semiconductor seemed to dominate its sensing traits. Fig. 11 [51] shows the field emission scanning electron microscopy (FESEM) image of un-doped and doped $SnO_2$ which clearly displays the decreased particle size and hence the increased surface area after V-doping.

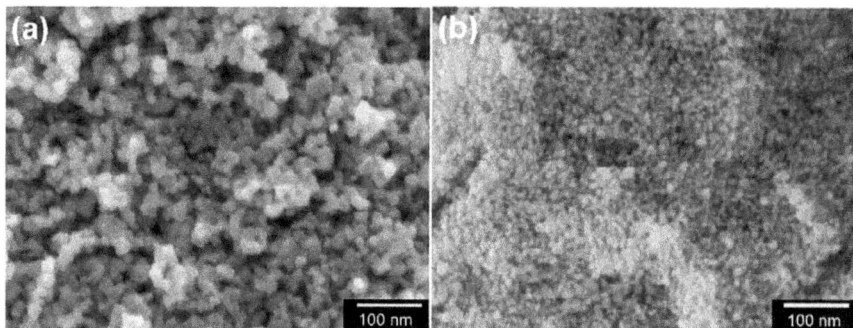

**Fig. 11.** *Images of FESEM analysis of (a) un-doped $SnO_2$ and (b) V-doped $SnO_2$. Reproduced here with the permission of Elsevier [51].*

The spectrometric analysis (XPS, FTIR) along with the combined XRD revealed that low vanadium doping was better for more crystalline $SnO_2$ & higher response to carbon monoxide gas. Moreover, CO response was in direct relation with CO concentrations in air. Mechanism of detection of CO by V-doped $SnO_2$ sensor has also been revised in which vanadium reduced cations $V^{3+}$, $V^{4+}$ acted as electron donors to facilitate $O_2$ adsorption and activation of the sensor surface.

In addition to vanadium metal doping, the semiconducting metal oxide, $SnO_2$ has also been altered by incorporating other metals like calcium, titanium [49], etc. for CO detection. Ghosh et al. [70] have detailed the sonochemical construction of calcium loaded $SnO_2$ sensor for fast recognition of minimal ppm levels of CO. Calcium has been found to play two functions based on the amount of loading; first, $Ca^{2+}$ ions replaced $Sn^{4+}$ from $SnO_2$ lattice altering the lattice parameters. This was the case when Ca loading was within the solubility limit. On the other hand, i.e., when it was greater than the solubility limit, calcium deposited at grain boundaries as calcium oxide inhibiting the growth of tin dioxide particles [71]. This sensor has been shown to detect as low as 1.0 ppm of CO in the atmosphere, with long term stability and selectivity to a larger extent.

In the process of advancement of carbon monoxide sensors, the factors that influence the sensing features to a greater extent include the size of metal oxide particles, type of structure at the microscopic level and the dopant metal characteristics [72, 73]. The particle size below 6 nm has been proved to dramatically increase the gas sensing properties [74]. The morphology of the sensor is also known to directly affect the reaction between the analyte gas & the sensor surface, and its diffusion rate [72, 75, 76]; the surface with more porosity being able to easily detect large gas molecules due to swift diffusion into the layer. As mentioned above, the type of metal additive is assumed to play a crucial role in determining the sensitivity of the sensor against the analyte gas. Inclusion of noble metals like Pd, Ag, Pt, Au has become an increasing practice [45, 48, 50, 77]. The special benefit of gold has been its ability to lower the operating temperature for CO oxidation [73, 77, 78].

Manjula et al. [73] have made use of PEG-6000 (polyethylene glycol) to synthesize nano-sized $SnO_2$ particles impregnated with gold, which led to greatly improved room temperature sensing of CO along with high selectivity and sensitivity at low concentrations. The use of PEG-6000 prevented the increase in the size of $SnO_2$ particles limiting the size to more or less around 3-5 nm. A hydrothermal procedure has been used to design the sensing material. The sensor at 1.5 wt.% of $Au/SnO_2$ displayed high response to even 10 ppm CO gas at operating temperature lower than 50°C with minimum interference from other gases, proving itself to be a promising device for CO detection.

The same hydrothermal method was used recently by Peng et al. [45] in designing of a 1.5 mol% $Pt-SnO_2$ nanocomposite based low temperature (80°C) CO sensor, which apart from being selective to carbon monoxide gas among a variety of atmospheric gases, exhibited excellent stability.

On the same lines, Wang et al. [67] have reported the synthesis of $Au/V-SnO_2$ nanostructure powder, which has been prepared by the co-precipitation method for low-temperature CO sensing. Here, gold and vanadium together on the $SnO_2$ matrix served to increase the stability and response of the sensor and decrease the times of response & recovery. Fig. 12 [67] shows the transmission electron microscopy (TEM) picture of $Au/V-SnO_2$ nanoparticles in which round Au particles are clearly labelled in the metal oxide nanostructure.

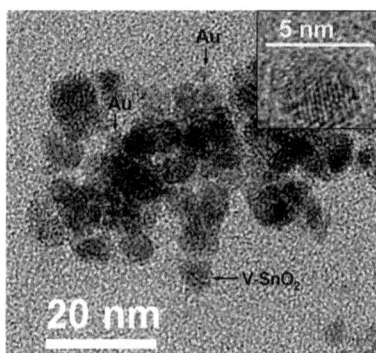

***Fig. 12.*** *TEM photograph of Gold/Vanadium Tin dioxide nanoparticle powder, HR-TEM image of gold particles (inset). Reproduced here with the permission of Elsevier [67].*

Au has been portrayed as the promoter in sensor surface activation by facilitating oxygen adsorption & inducing interaction between gold and vanadium cations.

Construction of $SnO_2$ devices in the form of a variety of nanostructures is being thoroughly explored. Apart from low dimensional nano-designs like rods, sheets, and cubes, more sophisticated (hollow & porous) $SnO_2$ structures like assembled flowers, cocoons, hollow spheres etc. are attracting more attention due to their high surface area applications. The fabrication of such a high-performance CO detecting device has been reported by Bing et al. [77] which was prepared by the multistep assembly of gold loaded tin oxide hollow multilayer nanosheets. Besides improvement in sensing traits by the inclusion of Au, the structural benefits involve hollow & multi-layered walls which could be easily penetrated, thereby facilitating the rate of adsorption of the gas analyte onto the sensor surface.

One of the most efficient methods for nanoparticle synthesis has been the reverse micelle (RM) formation, which takes place by mixing aqueous surfactant solution with organic solvent in an appropriate ratio. This method results in the formation of water droplets of size 1-10 nm dispersed in organic media. This procedure has been exploited for the manufacturing of a variety of substances like metals [72, 79], metal oxides [80], supported catalysts [81], as well as nanocomposites [82]. The advantage of this technique is the concoction of nano-sized particles as reaction place which can further be deposited by other nanoparticles in a homogenous fashion. Yuasa et al. [72] have used the RM procedure to fashion PdO-loaded $SnO_2$ nanoparticles, calcinated them and then prepared thick film type sensors for CO recognition. Fig. 13 [72] depicts the schematic procedure for PdO-loaded $SnO_2$ nanoparticle preparation by the RM method.

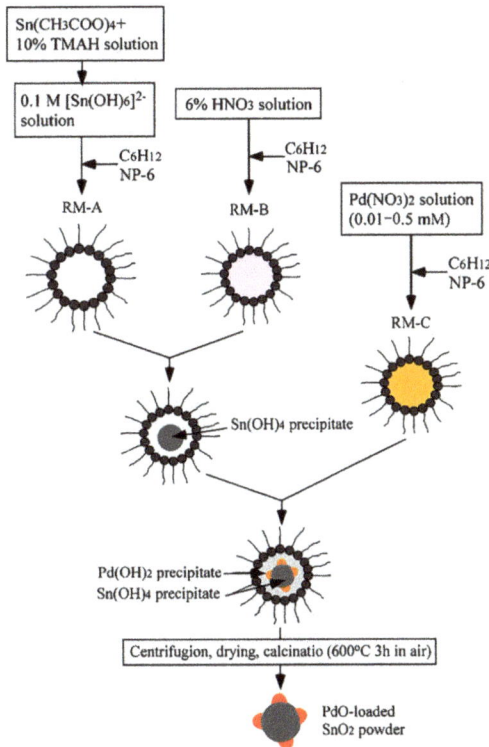

***Fig. 13****. Diagrammatic representation of the reverse micelle procedure for the synthesis of PdO-loaded SnO₂ nanoparticles. Reproduced here with the permission of Elsevier [72].*

The single crystalline nature of the $SnO_2$ particles was confirmed using TEM and high-resolution TEM. The data indicated the high thermal stability of $SnO_2$ nanoparticles. The 0.1% PdO loaded $SnO_2$ generated maximum response towards CO sensing.

More exploration in the domain of sensitive CO sensors led to the evolution of devices capable of sensing very low concentrations of atmospheric carbon monoxide at low/room temperature conditions. Kim et al. [48] have devised a Pd/$SnO_2$ nanoparticle-based CO sensor, which has been observed to be competent for sensing 6-18 ppm of CO concentrations at low temperature (60°C) with remarkable stability & high sensitivity in response. The sensor once exhausted was also reported to be recoverable by slight application of heat. A similar effort has been made by Wang et al. [50] for high-performance room temperature sensing of carbon monoxide gas using platinum loaded tin dioxide porous nano-solid (PNS) represented as $SnO_2$: Pt PNS. The loading of Pt on $SnO_2$ resulted in a decrease of the depletion region and hence an increase in the electron conduction channel when brought in contact with CO, via both electric and chemical mechanisms (Fig. 14) [50]. The sensor thus displayed huge potential for effective CO sensing applications.

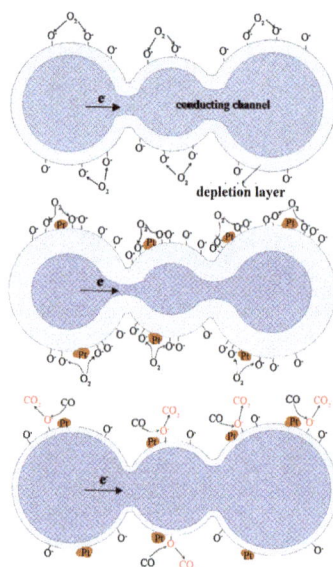

***Fig. 14****.Diagram explaining the increase in sensing traits by Pt- loading. (i)$SnO_2$ PNS sensor response in air, (ii)$SnO_2$: Pt PNS sensor response in air, (iii) $SnO_2$: Pt PNS sensor response in CO atmosphere. Reproduced here with the permission of Elsevier [50].*

A recent study has been published by Debataraja et al. [46] on nanostructure composite involving thick film palladium doped graphene-$SnO_2$ sensor for carbon monoxide recognition purposes. The sensor gave effective signals with 0.1% Pd doping even towards 1.0 ppm CO, revealing its capability to be utilized as a sensitive detector.

**(b) Zinc oxide (ZnO) based sensors**

Zinc oxide (ZnO) is also an *n*-type semiconductor with a large bandgap of 3.37eV, high thermal stability, a large value of mobility of conduction electrons, non-toxic nature, low cost and recyclability [83-87]. Owing to these special features, it has been tremendously utilized for gas sensing objectives. As per literature, copious experiments have been performed in order to develop stable, sensitive, selective & efficient ZnO based sensors for CO, one of which includes their framing into a variety of nanocrystallites for enhanced sensing [22, 88]. Recent discoveries will be presented in somewhat detail in the following part of the chapter.

One of the findings includes the synthesis of flower-like, single-crystalline nanorod assemblies of ZnO proposed by Prabhakar et al. [89] for the detection of small traces of carbon monoxide gas (Fig. 15) [89].

***Fig. 15**. SEM pictures of ZnO floral assemblies (a), (b) Fl-1 (thin rods) and (c),(d) Fl-2 (thick rods). Reproduced here with the permission of Elsevier [89].*

Out of the two differently sized flower-like rods, only the thinner ones were able to show sensitivity towards CO gas (~10 ppm) at 200°C, which has been attributed to the availability of larger surface atoms in thinner rods that led to an increased adsorption activity. Prajapati et al. [38] have exploited the same idea of increasing the surface to volume ratio for improved adsorption-desorption mechanisms of analyte gas at the sensor surface. An immensely selective and sensitive honeycomb type zinc oxide nanostructure (Fig. 16) [38] for CO recognition has been reported to be developed through the study.

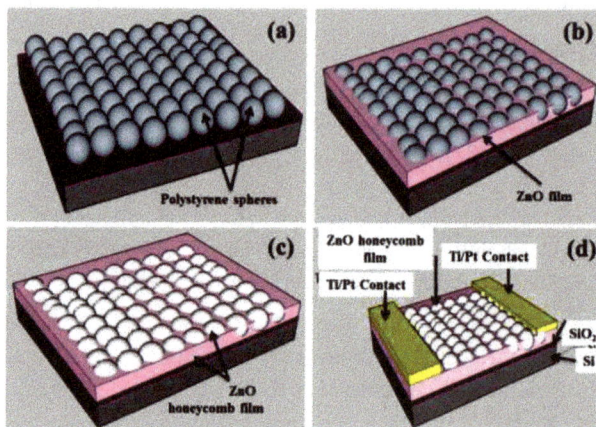

**Fig. 16.** *Pictorial representation of the synthesis of honeycomb-ZnO nanofilm: (a) Self-assembly of polystyrene spheres on SiO₂/Si film, (b) ZnO thin deposition on the spheres, (c) ZnO honeycomb nanostructure left behind after lift-off of the spheres, (d) Metal contacts for gas sensing purpose. Reproduced here with the permission of Elsevier [38].*

The sensor manufactured by colloidal lithography & lift-off procedure has been proposed to be cost-effective as well. It can detect up to 500 ppb concentration of CO with a maximum response of ~81.2% for 3 ppm CO at 300°C. Mixed metal oxides have also been considered during the study, $CeO_2$-ZnO mixed films [84] being one of the examples. A tremendous improvement in the detection limit has been achieved by Choi et al. [30] who reported the sub-ppm level (100 ppb–2 ppm) room temperature detection of CO through Pd nano-dots functionalized surfaces of selectively grown ZnO nanowire networks. The surface of networked ZnO nanowires was functionalized by palladium via γ-ray radiolysis.

Parallel research revealed an enormous refinement in the detection limit for CO sensing by functionalization of the novel $SnO_2$-ZnO core-shell nanowires with gold nanoparticles [90]. Fig. 17 [90] shows the step by step preparation of the sensor. The sensor generated a very high response to 100 ppb concentration of CO gas, with minimum cross-sensitivity from other gases.

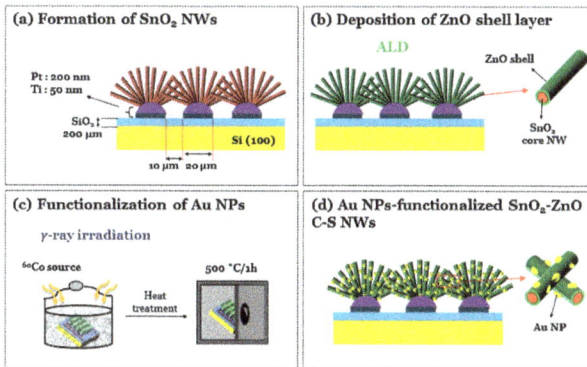

**Fig. 17.** *Scheme of synthesis of Au functionalized SnO$_2$-ZnO nanowires (a) Preparation of networked SnO$_2$ nanowires by vapour-liquid-solid method, (b) Coating of ZnO shell on SnO$_2$ core using atomic layer deposition, (c) Attachment of Au nanoparticles on the nanowires using γ-irradiation, (d) Resultant Au functionalized SnO$_2$-ZnO nanowire-based sensor for CO. Reproduced here with the permission of Elsevier [90].*

As was the case with tin dioxide sensors, metal inclusions (e.g. Al [22], Au [83, 85], etc.) upgraded the CO recognition limits in ZnO based devices also, attaining sub-ppm level sensitivity.

Furthermore, 3D reduced graphene oxide (RGO) [91] & multiwalled carbon nanotubes (MWCNT) [92] have been separately used in conjunction with zinc oxide thin films to investigate the selectivity and sensitivity of the sensor response.

### (c) Indium(III) oxide (In$_2$O$_3$)and other miscellaneous metal oxide-based sensors

Indium (III) oxide (In$_2$O$_3$) has also been deployed for sensing objectives, which proved to be an encouraging substitute for most widely exploited stannic oxide (SnO$_2$) & ZnO metal oxide semiconductors. It has often been used with transition metal dopings like Zn [23], Co [93], Au [94], etc. to affect room temperature sensing of low concentration of CO in most cases.

Although barium stannate (BaSnO$_3$) [95], cerium oxide [63] & molybdenum trioxide (MoO$_3$) [96] thin films were individually used as CO detectors, detection limits have not been convincing at low/room temperature. Cobalt oxides, such as Co$_3$O$_4$ spinel nanostructures [97] & cobalt oxyhydroxide (CoOOH) [98] have also been investigated for their carbon monoxide sensing capabilities and were found to be competent enough to notice the existence of low ppm concentrations of CO.

Some $p$-type semiconducting oxides like ferric oxide ($\alpha$-Fe$_2$O$_3$) nanoparticles loaded on layered reduced graphene oxide nanosheets have been studied by Basu et al.[90]for their fast, sensitive and selective response to about 10 ppm CO room temperature [90, 99]. Fig. 18 [90] illustrates the mechanism of gas sensing by the $p$-type semiconductor.

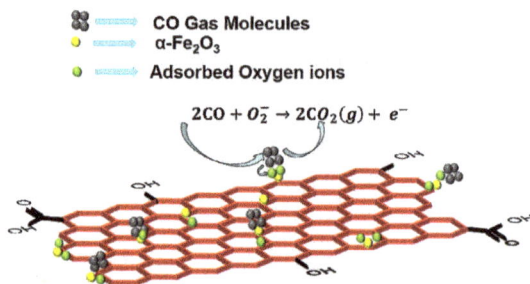

**Fig. 18.** *Mechanism of sensing of reduced graphene oxide-$\alpha$-Fe$_2$O$_3$ p-type semiconductor. Reproduced here with the permission of Elsevier [99].*

Recently, the less studied rare earth metal oxide, samarium(III) oxide (Sm$_2$O$_3$) has been synthesized as nanorods using the hydrothermal procedure and analyzed for its gas sensing features, which displayed significantly good potential to be deployed for CO detection [99]. The limit of detection has been found to be 1.0 ppm for carbon monoxide.

Another recent study by Kumar et al. [100] has revealed that octahedral molecular sieves (OMS-2) nanofibers composed of manganese oxide can also be utilized for sensing applications because of their large surface area for adsorption, high stability and specific catalytic activity. Doping of metals like Ni, Cu, Co, Ag, Fe, Nb, etc. are known to increase the efficiency of OMS-2 nanofibers towards sensing [101]. Kumar et al. [100] have prepared Nb-doped K-OMS-2 for room temperature CO detection. The response of the sensor (S) was determined by the relation shown in equation 1 [100].

$$S = \frac{R_a - R}{R_a} \; X \; 100 \tag{1}$$

where $R_a$= Resistance of Nb-OMS-2 sensor in the air in the absence of CO

R = Resistance of the sensor in the presence of CO

The sensor sensitivity and selectivity towards CO gas in the presence of other interfering air pollutants have been investigated in detail. The sensor films were exposed to different concentrations of carbon monoxide gas, which revealed that the sensor can detect 2.0 ppm of carbon monoxide gas by displaying an (S =) 22.35% response. An increase in response to 33% (at 10 ppm CO), 51% (at 300 ppm CO), 55.12% (at 100 ppm CO), 75% (at 200 ppm CO) and 91% (at 400 ppm CO) has been observed using an increasing concentration of carbon monoxide gas. Average response and recovery times between 25-40 s have been observed throughout the study. A similar response and recovery times have been observed up to 400 ppm carbon monoxide concentration. In addition, a linear response towards carbon dioxide was displayed by the sensor film up to 200 ppm, which may also be considered linear up to a CO concentration of 400 ppm.

A mechanism of CO sensing has been devised by using the Nb-doped K-OMS-2 film. As per mechanism, the adsorption of CO occurred at Nb sites, which can be explained on the basis of energy and area considerations. The CO reacted with bridge surface oxygen to get oxidized to $CO_2$ that lower the resistance response of the sensor. The recovery of the sensor was demonstrated in the last step using atmospheric oxygen, which helped in the replenishment of the vacancies. The sensor nanofibers were sensitive to even 2.0 ppm of CO with sufficiently high response and stability.

**(ii) Carbon nanotubes as carbon monoxide gas sensors**

Carbon nanotubes have been employed as sensors for CO gas [102]. Carbon nanotubes are similar to graphite in terms of hybridization and bonding among carbon atoms [103]. Single-walled (SWCNTs) & Multi-walled carbon nanotubes (MWCNTs) are two of their types. SWCNTs are monoatomic layered $sp^2$-bonded carbon cylinders in which atoms are arranged in a honeycomb-like structure (Fig. 19) [104, 105].

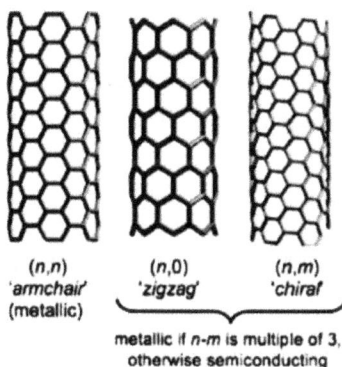

*Fig. 19. Pictorial representation of single-walled carbon nanotubes: armchair nanotube, Zig-Zag nanotube, and Chiral nanotube. Reproduced here with the permission of IOP Publishing, Ltd [104].*

Multi-walled carbon nanotubes comprise of concentrically arranged SWCNTs like rings in a trunk [106]. The nanotubes are hollow along the nanotube axis & their physical properties are known to depend on the nanotube diameter [105]. SWCNTs of small-diameter are stiff and strong with a high value of Young's modulus having high tensile strength [106]. Carbon nanotubes, in general, exhibit very high thermal conductivities at room temperature [107] and even superconductivity have been observed in SWCNTs at low temperature [108].

CNTs generally behave like $p$-type semiconductors. The property of the nanotubes responsible for their application in sensing devices is their enormous adsorption capacity because of the very large surface area [102]. Joseph et al. [102] have devised a method regarding MWCNTs utility as a high-temperature CO sensor, which gave an optimal response to CO at 600°C.

More recent research has unveiled the fabrication of single-walled carbon nanotubes co-functionalized by $SnO_2$ & Pt nanoparticles and then studied its application in CO recognition [109]. The composite nanostructure prepared has been found to be capable of detecting even 0.05 ppm CO at room temperature, which made it viable to be used in this field.

**(iii) Conducting polymeric films-based CO gas sensors**

Sensors based on conducting polymeric films have been presented as a finer choice in comparison to metal oxide semiconductor-based sensors, advocating some special features like low cost, easy packing, remote positioning ability, reusability and stability of sensor besides high sensitivity, selectivity and rapidity of the response [110].

The conductivity of the polymers can be managed as a function of level and type of doping i.e., their efficiency is increased manifolds when these are doped with appropriate metals. Polyanilines, polythiophenes & polypyrroles are some of the examples of conductive polymers [111]. The polymeric sensor films respond to the target gas by changing their resistance value.

Misra et al. [110] have put forward the synthesis of a nanocrystalline thin film of Fe-Al doped polyaniline via a vacuum deposition technique. Its characterization and gas sensing tests have been reported. A comparison of UV spectra of doped and undoped polyaniline indicated upon the changes in the conjugation caused due to Fe-Al doping. The vacuum deposition procedure led to the formation of crystallites in polyaniline film as opposed to the amorphous nature of polymers in general. The gas detection has been accomplished by examining changes in current-voltage attributes of the polymeric film. Fig. 20 [110] represents the design of the sensor.

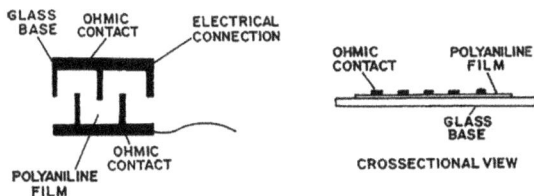

***Fig. 20.*** *Diagram of design of the sensor. Reproduced here with the permission of Elsevier [110].*

Polyaniline being a $p$-type semiconductor conducted via hopping movement of polarons and dipolarons through intercrystallite boundaries, where charge transfer resulted in the development of charge barrier. This barrier got reduced in response to the adsorption of CO gas leading to a hike in output current response of the sensor.

The sensor has been identified as fairly sensitive towards low concentrations of carbon monoxide gas (0.02-30 ppm) at room temperature. Moreover, it has been claimed to be reusable with a longer shelf life [110].

Comparable works have been published involving the functionalization of polypyrrole with electroactive species like ferrocene [112] or iron porphyrins [113] during polymerization (Fig. 21 [112] and 22 [113]). Such materials have been found to exhibit room temperature sensing of CO with reasonably good response & recovery characteristics.

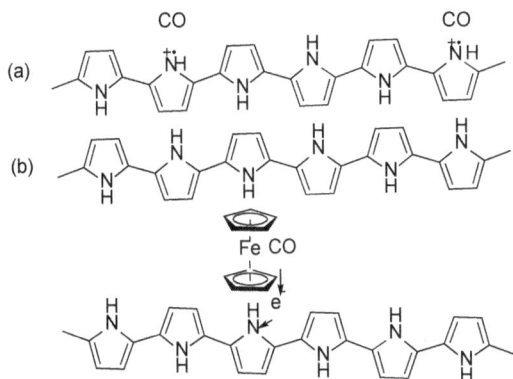

***Fig. 21.*** *Pictorial representation of the interaction of CO with (a) Polypyrrole, (b) Ferrocene doped polypyrrole [112].*

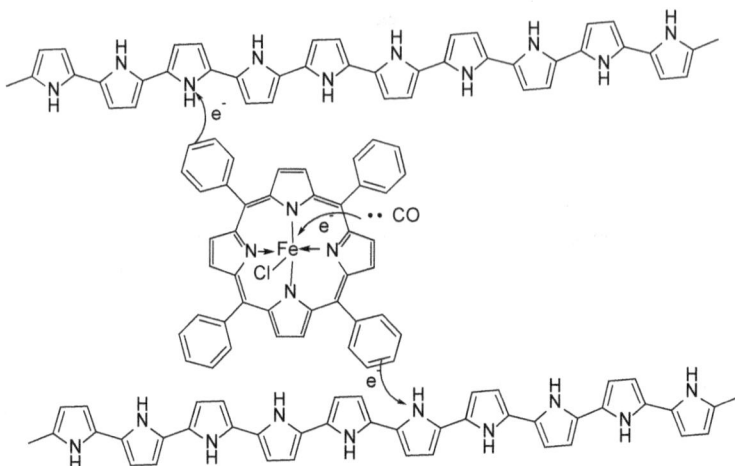

***Fig. 22.*** *Diagrammatic scheme of interaction of Fe-porphyrin functionalized polypyrrole with CO [113].*

## (iv) Colorimetric sensors for CO gas

Colorimetric detection refers to a method of sensing in which interaction of the sensor with the analyte results in a visible colour change. The change in colour can also be monitored through UV-visible spectroscopy. Although chromogenic systems capable of recognizing CO are not abundant but this branch of CO sensors is a matter of ongoing research. Most widely reported detectors include complexes of Rh [114], Ru [115] & Os, etc.

Esteban et al. [116] have unveiled a selective and sensitive sensor model involving the binuclear rhodium complex as the moiety to interact with CO gas. The metal complex reaction with CO involved two steps: first, axial coordination of CO and second, the back donation of electrons from the metal to $\Pi^*$ of CO. The rhodium complex selected for the study was a dirhodium compound prepared by Cotton et al. [117] in 1985.Fig. 23 [116] outlines the reaction between CO and the sensing component.

A novel idea of adsorption of a compound on silica gel has also been devised, which led to the development of grey-violet colored silica. The colored silica upon exposure to the different concentrations of CO produced a sharp colour change to orange-yellow within a few minutes (Fig. 24) [116].

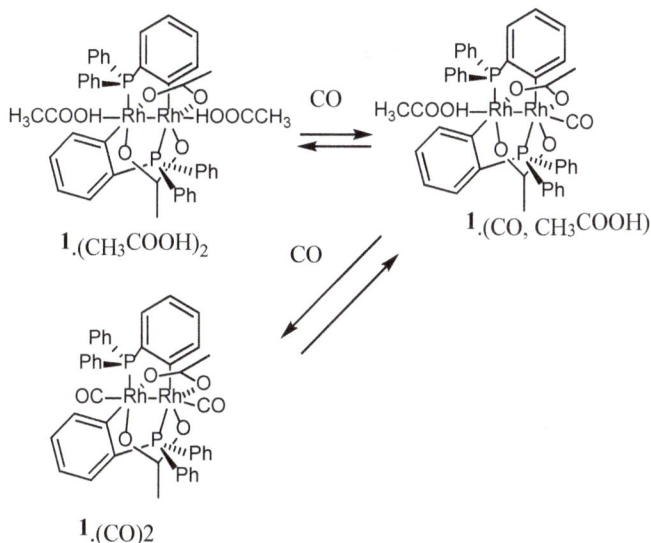

*Fig. 23. The reaction of sensor1·(CH₃CO₂H)₂with CO to result in its mono and di co-ordination at axial positions. Reproduced here with the permission of John Wiley and Sons [116].*

*Fig. 24. Pictures of compound 1(CH₃CO₂H)₂ adsorbed on silica gel in absence of CO (left), and in presence of 50 ppm of CO (right). Reproduced here with the permission of John Wiley and Sons [116].*

Additionally, the sensing material was noted to be selective and reversible. The same complex with trifluoroacetate in place of acetate ligands has been deployed for the same purpose by Courbat et al. [118].

Lin et al. [119] have extended the scope of CO monitoring by introducing an authentic, easy to operate, miniaturized device based on colorimetric detection by a specific chemical probe. It has been designed for personal exposure tracking purposes. The chemical reagent used for sensing CO was $K_2Pd(SO_3)_2$, which was immobilized on porous silica gel during the fabrication of sensor chip. The chip components have been described in Fig. 25 [119].

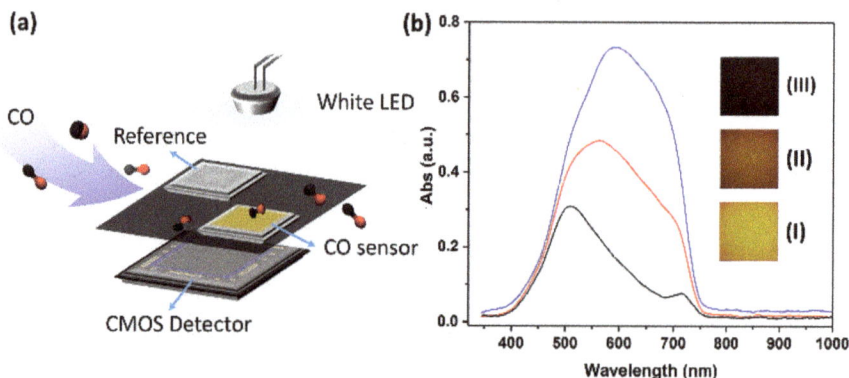

**Fig. 25**. *Representation of (a) sensor design comprising of grey-reference chip and yellow-CO sensor chip. The sensor chip is illuminated by a white LED and the transmission signal is detected by the CMOS detector. (b) UV-Vis spectrum of the sensing component in presence of increasing exposure to CO. (I), (II), (III) insets display the photographs of CO sensor chips after exposure to different concentrations of CO gas from low to high. Reproduced here with the permission of ACS [119].*

A sensing mechanism has been devised, which explained the reaction of $K_2Pd(SO_3)_2$ with CO on silica (Scheme 1) [119].

$$CO + K_2Pd(SO_3)_2 \longrightarrow \quad _2 \quad _2 \quad _2SO_3$$

*Scheme 1. The reaction involved between carbon monoxide and $K_2Pd(SO_3)_2$ [119].*

The sensor has been observed to achieve sensitivity towards a wide range of CO concentrations with an adequate response towards 0-10ppm CO (Fig. 26) [119].

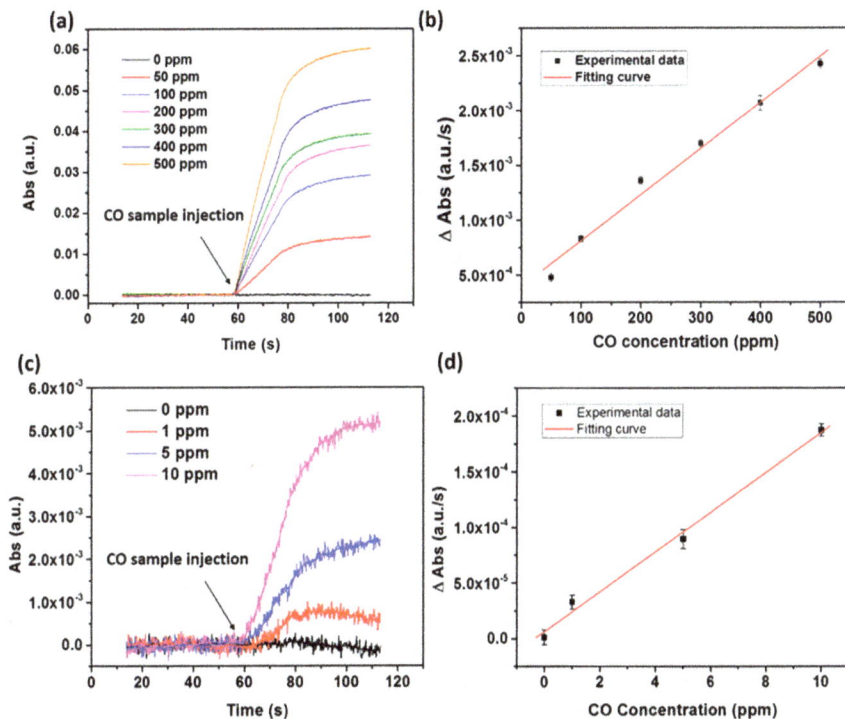

***Fig. 26***. *(a) Responses of the CO sensor to different concentrations of dry CO gas samples (after baseline correction). (b) Sensor response calibration with respect to CO. Absorbance change (ΔAbs (au/s)) was calculated from slope difference between baseline and sampling periods (R₂ = 0.9822, slope = (4.202 ± 0.252) × 10⁻⁶ au/s·ppm). Error bars represent the standard deviation. (c) Responses of the CO sensor to lower concentrations of dry CO gas samples (after baseline correction). (d) Sensor response calibration with respect to CO (R₂ = 0.9889, slope = (1.7879 ± 0.152) × 10⁻⁵ au/s·ppm). Reproduced here with the permission of ACS [119].*

The detection limit was measured to be satisfactorily low (1.0 ppm), with fast response time (20s) and negligible interference from other air pollutants (Fig. 27) [119]. High sensitivity and selectivity eliminated the need for any masking agent for potential interferents. The sensor can colorimetry detect the sub-ppm level of carbon monoxide with sufficiently long sampling time. The sensing system can also be programmed to adjust to sampling time to achieve better sensitivity with a low concentration of an analyte. Moreover, the sensor chip was professed to have the ability of getting functionalized with different chemical reagents, depending on the specific analyte to be sensed. The sampling time can be adjusted to a short duration while dealing with a high concentration of an analyte. The sensor can be used for continuous monitoring of CO concentration also implies the continuous use of a single sensing chip. The study indicated that the use of a sensing chip continuous for 8h led to only a 10% reduction in sensor response towards carbon monoxide gas. The sensor displayed a response better than the WHO prescribed limit (10 ppm, 8 h mean) for CO concentration in air [119].

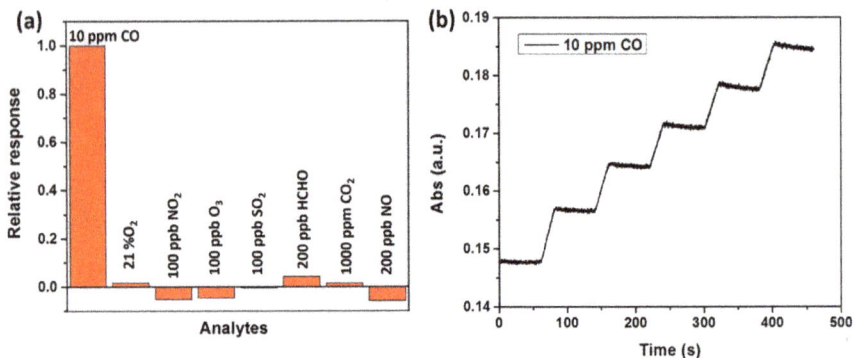

***Fig. 27.*** *(a) The interference study indicating the effect of common gaseous species (21% $O_2$, 100 ppb $NO_2$, 100 ppb $O_3$, 100 ppb $SO_2$, 200 ppb HCHO, 1000 ppm carbon dioxide, 200 ppb NO) during the detection of carbon monoxide. Relative response has been calculated by taking a ratio of carbon monoxide sensor response to the interferents and carbon monoxide sensor response towards a 10 ppm concentration of CO. (b) The carbon monoxide response displayed during continuous monitoring of CO (10 ppm) at 75% RH (60 s purging time and 20s sampling time). The change in sensor response for different sampling periods was observed to be less than 6%. Reproduced here with the permission of ACS [119].*

Ruthenium(II) and osmium(II) vinyl complexes have also been investigated for their sensitivity towards CO gas by Toscani et al. [120]. Fig. 28 [120] lists all the ligands used in Ru or Os complexes and Fig. 29 [120] depicts the reaction between the chemical probe and CO.

M = Ru (**2**)
M = Ox (**3**)      R =

M = Ru (**4**)      R =
M = Os (**5**)

M = Ru (**6**)      R =

M = Ru (**7**)      R =

M = Ru (**8**)      R =

M = Ru (**9**)      R =

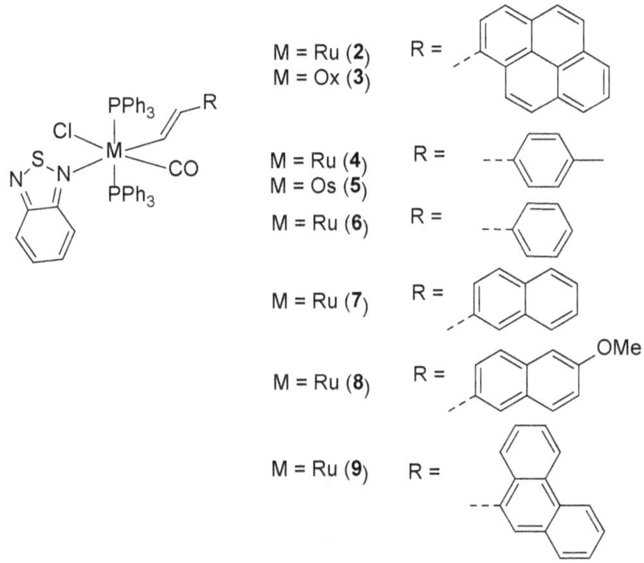

**Fig. 28**. *List of Ru and Os complexes 2-9[120].*

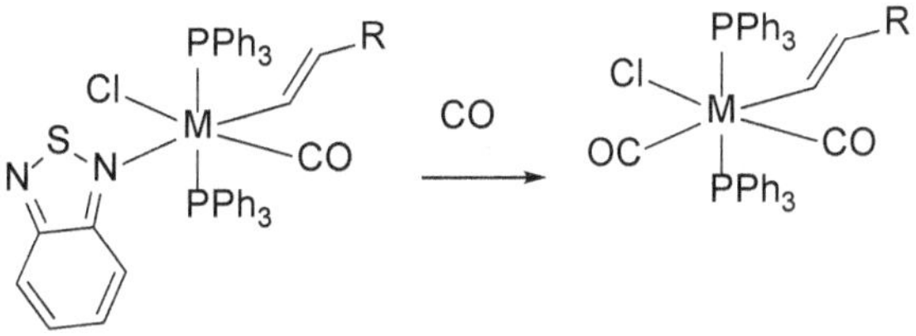

**Fig. 29**. *The reaction of complex (PR) with CO to form dicarbonyl complex (PR-CO)*
*[120].*

CO gas sensing has been accomplished by the same procedure of adsorbing the compound on silica, then its exposure to CO concentration in air. Fig. 30 [120] illustrates the colour changes of different probes on interaction with different concentrations of CO.

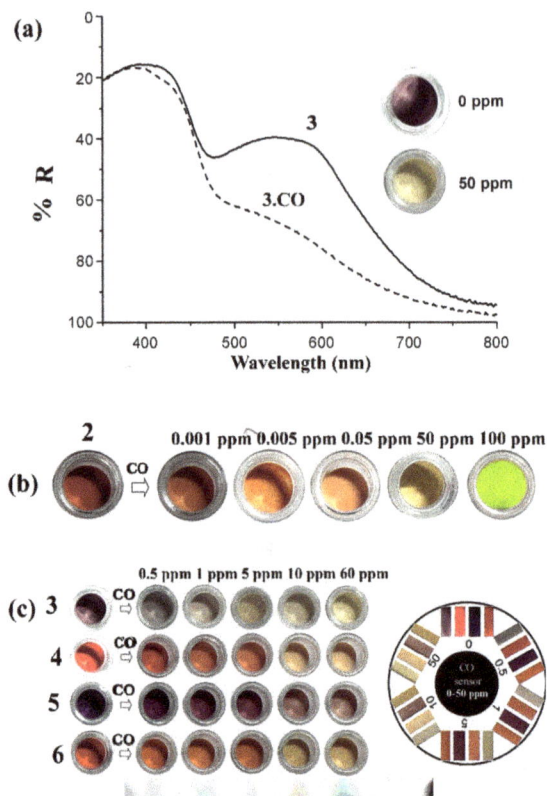

***Fig. 30.*** *(a) Diffuse-reflectance UV/Vis spectrum of complex 2 on silica depicting the response in absence and presence of 50ppm of CO. b) Colour changes observed for complex 1 on silica gel upon exposure to various concentrations of CO in air. c) Sensing array of probes 2–5 for CO detection in the air that showed visible colour changes upon exposure to 0.5, 1, 5, 10, and 50 ppm CO. Reproduced here with the permission of John Wiley and Sons [120].*

All the compounds yielded selective response to low concentrations of CO.

Except ruthenium complexes, naphthalimide based probes involving palladium metal have been exploited for fluorescent CO detection in the air by Feng et al. [121].

## (v) Non-dispersive infrared (NDIR) principle-based CO sensor

According to the Non-dispersive infrared (NDIR) principle, particular gases absorb characteristic wavelengths in the infrared spectrum, which can be exploited for the identification of gases [122]. NDIR based gas sensor constitutes an IR source, a gas chamber and a detector (Fig. 31) [122].

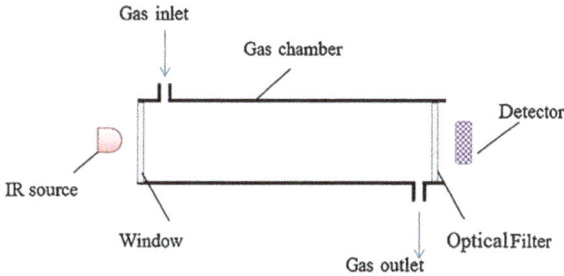

**Fig. 31.** *Structural design of NDIR principle-based gas sensor. Reproduced here with the permission of Elsevier [122].*

NDIR principle of gas sensing is based on Beer-Lambert's law (Equation 2) [122]:

$$I = I_o\, e^{-kCL} \tag{2}$$

where $I_o$ = Initial beam intensity of radiation

I = Beam intensity after absorption by the gas

k = absorption coefficient

C = gas concentration

L = sample chamber length

Low energy consumption is one of the advantages of NDIR spectroscopy in comparison to other spectroscopic techniques. Diharja et al. [123]have devised a CO gas sensor based on the NDIR principle. The incandescent light bulb has been utilized as the IR source for the experiment. An optical bandpass filter has been attached to transmit a specific portion of light, characteristic of CO. The detector used for the detection of light after absorption was TPS 334 thermopile. Fig. 32 [123] and 33 [123] detail the NDIR sensor equipment and the working diagram of the sensor system respectively.

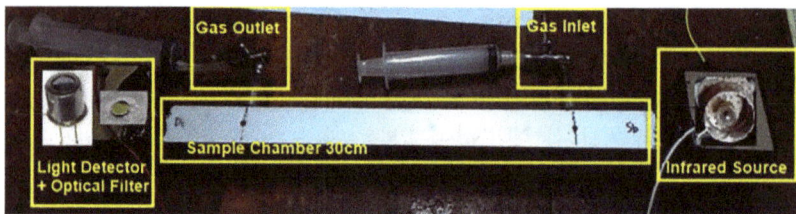

***Fig. 32****. Picture of the NDIR sensor displaying the constituent equipment [123].*

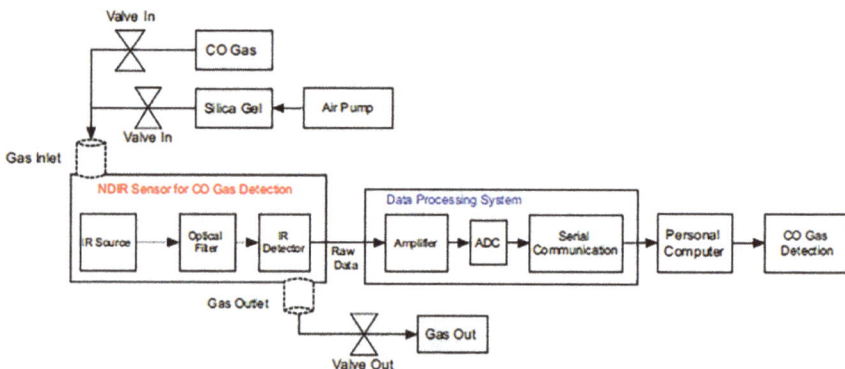

***Fig. 33****. Schematic diagram of the NDIR sensor working [123].*

The dependence of NDIR sensor voltage (mV) on CO gas concentration (ppm) was expressed as shown in equation 3 [123]:

$$y = 7.135 \ln(x) - 1.573 \tag{3}$$

Fig. 34 [123] shows the graphical relationship between thermopile voltage response and CO concentration after baseline correction.

The sensitivity of the NDIR sensor was calculated as 7.0 mV/ppm.

**(vi) Electrochemical gas sensors**

Electrochemical gas sensors are based on the principle of measurement of the electrical signal generated as a result of a chemical reaction between electrode and analyte gas [124]. One of the important components of electrochemical gas sensors is the set of electrodes, which can be two (sensing electrode, reference electrode) or three (sensing electrode, a reference electrode, and counter electrode) in number as per requirement. To receive a substantial signal for gas sensing, the electrodes must have a large surface area. The sensing electrode (also known as working or active electrode) is generally composed of noble metals like Au, Pt etc. or it can be made of a material coated with Pt, Pd or carbon. It should also include a hydrophobic porous polymeric material in its composition, which can act as a connecting platform between the atmosphere and the working electrode, for the purpose of gas diffusion to the sensor [125].

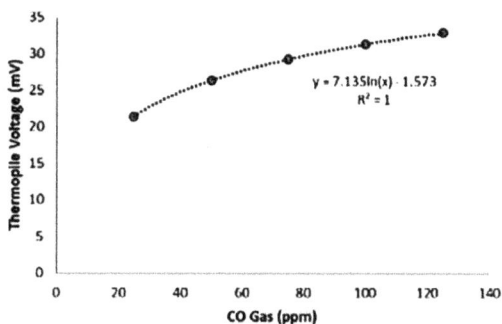

**Fig. 34.** *Baseline corrected voltage response of NDIR to CO concentrations [123].*

The electrochemical sensors are extensively deployed for gas detection purposes due to their additional advantages like:

i. The sensor is designed such that the output signal is linearly proportional to gas concentration, which allows better precision and easier calibration [125].
ii. Lowest power consumption among other sensor devices [124].

The electrical signal measured by the electrochemical sensor can be current, potential, conductivity or capacitance, on the basis of which the sensors can be divided into various types, while only the following major types will be discussed here[124, 126]:

a. Potentiometric sensors
b. Voltammetric sensors (Amperometric)

In the first type, i.e. Potentiometric sensors, an equilibrium potential difference occurring between sensing and the reference electrode is measured [125]. No current flow takes place at equilibrium in an ideal situation. The potential difference determined, depends on the concentration of the electroactive species by the Nernst equation.

The second type i.e. Voltammetric sensors includes the measurement of current-voltage variation [125]. Amperometry is a special case of voltammetry where the constant potential is applied to the sensor, and current response (proportional to electroactive species concentration) is monitored.

## (a) Potentiometric sensors

Potentiometric gas sensors are mostly solid electrolyte based, which can be categorized into three types [127, 128].

*Type A* sensors involve the direct participation of mobile ions (of solid electrolyte) in the equilibrium established with chemical species responsible for fixing the potential [127, 128]. Type A is further divided into two subcategories based on types of electrodes used.

I. The first subcategory sensors use the electrodes in which only one chemical potential of a neutral constituent is required. Fig. 35 [128] represents the potentiometric oxygen gas sensor which is an illustration of Type A- subcategory I sensors.

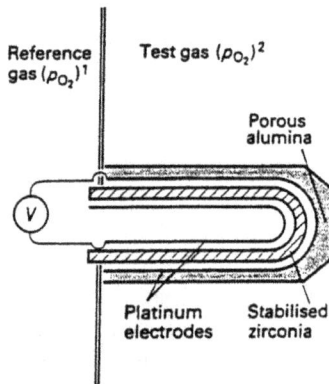

*Fig. 35. Diagrammatic representation of solid-state potentiometric oxygen gas sensor – an example of Type A subcategory I sensor. Reproduced here with the permission of IOP Publishing [128].*

The reaction involved at cathode and anode is depicted in scheme 2 [128].

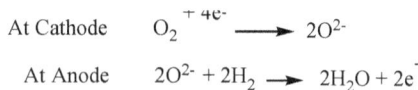

$$\text{At Cathode} \qquad O_2 \xrightarrow{+\ 4e^-} 2O^{2-}$$

$$\text{At Anode} \qquad 2O^{2-} + 2H_2 \longrightarrow 2H_2O + 2e^-$$

*Scheme 2. The reaction at electrodes [128].*

The associated equation to analyze the reaction shown in scheme 2 is given in equation 4.

$$V = \frac{RT}{4F} \ln\frac{(p_{O_2})^1}{(p_{O_2})^2}$$ (4)

where superscript 1: reference gas – air

superscript 2: test gas – the atmosphere in automobile exhaust

The second subcategory of Type A potentiometric sensors uses the electrodes which involve coupled equilibria between species to be detected and the mobile ions. An example of the cell is:

*Na/Na⁺ beta alumina/ Na₂CO₃/ CO₂(g), O₂(g)*

The reaction at the sensing electrode is shown in Scheme 3[129, 130].

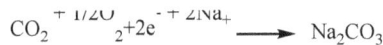

$$CO_2 + 1/2O_2 + 2e^- + 2Na^+ \longrightarrow Na_2CO_3$$

*Scheme 3. The reaction at the sensing electrode [129, 130].*

*Type B* potentiometric gas sensors involve electrolytes in which mobile ion is not in equilibrium with the species to be detected [127, 128]. Therefore, in order to equilibrate solid electrolyte with the atmosphere, an ion related to the target gas is inculcated into the solid electrolyte structure.

*Type C* potentiometric solid-state gas sensors involve multiple reactions taking place at the electrodes resulting in the generation of "mixed potential" [127, 128]. Almost all the potentiometric sensors discussed below belong to this category.

Stabilized zirconia has widely been used as a solid electrolyte material in potentiometric solid-state gas sensors, which is defined as an oxide ion conductive substance [131, 132].

Okomata et al. [132] have formulated a new CO gas sensor making use of stabilized zirconia ($ZrO_2$) in combination with two electrodes, one on each side. This sensor did not require a reference $O_2$ gas for its functioning. Out of the two electrodes, one was Pt electrode in direct contact with sample gas and the other was a Pt pseudo-air electrode coated with a CO oxidation catalyst, which was not kept in direct contact of the sample gas (Fig. 36) [132].

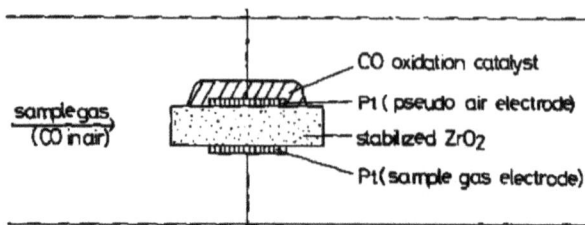

**Fig. 36.** *Diagram of CO gas sensor. Reproduced here with the permission of Elsevier [132].*

The bare Pt electrode gave anomalous electrode potential. The coated Pt electrode contained only $O_2$ and a small amount of $CO_2$ at its surface, which was formed due to the reaction of a small amount of CO with $O_2$ (in the air) during diffusion through the porous coating to the electrode. Hence, the covered Pt electrode gave pseudo-air electrode potential. The sensor was thus capable of delivering emf responses corresponding to CO partial pressures without the need for reference $O_2$ gas.

The CO oxidation catalyst (Alumina supported on Pt electrode) played a vital role in the sensor assembly. The sensor can be used for the detection of low concentrations of CO at 300°C (Fig. 37) [132].

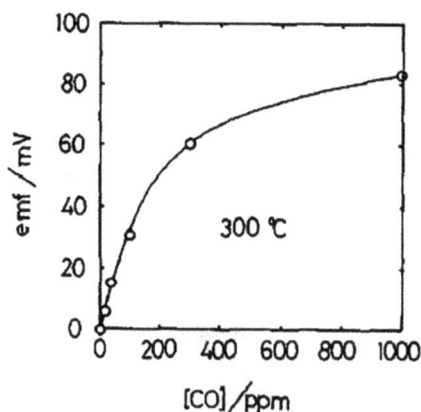

**Fig. 37.** *Sensor emf response towards different concentrations of CO [132].*

The emf of the sensor has been assumed to be generated form mixed electrode potential arising from multiple electrochemical reactions of $O^{2-}$ with CO, and adsorption of oxygen during CO oxidation reaction on Pt [132]. Another CO potentiometric gas sensor generating mixed electrode potential has been recently reported by Yang et al. [133]. The authors have employed a YSZ (yttria-stabilized zirconia) as the solid electrolyte in the sensor model along with $NiFe_2O_4$ and Pt as the sensing and the reference electrodes respectively (Fig. 38) [133].

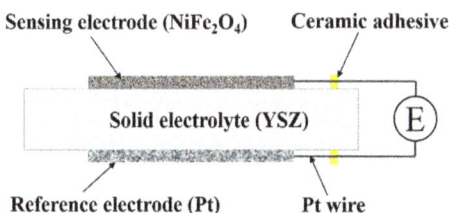

**Fig. 38**. *Pictorial representation of the potentiometric solid-state sensor. Reproduced here with the permission of the Electrochemical Society [133].*

The mixed potential response of the sensor displayed linear dependence on logarithmic CO concentration (Fig. 39) [133].

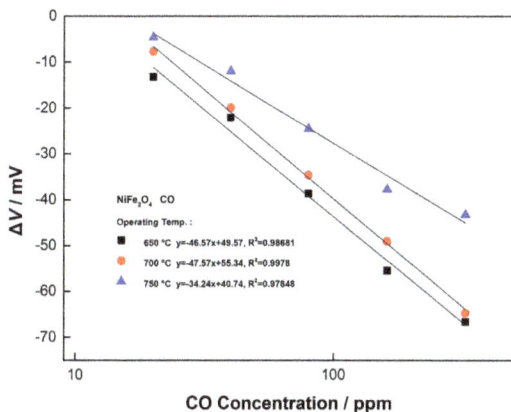

**Fig. 39**. *A plot of sensor mixed potential response with the logarithm of CO concentration at different temperatures of operation. Reproduced here with the permission of the Electrochemical Society [133].*

The use of $NiFe_2O_4$ enhanced the response signals and CO selectivity. A reduction in the response & recovery times along with the limit of detection (LOD) to 20ppm has also been observed. The temperature of the operation of the sensor was between 650-750°C with maximum sensitivity at 700°C. It has also been demonstrated to possess long term stability [133].

Stabilized zirconia based sensors have been used with different electrodes for CO sensing purposes. Miura et al. [134] have utilized a combination of oxides namely, CdO and $SnO_2$ as the electrode couple. The sensor has successfully detected CO concentrations ranging from 20-4000 ppm yielding a linear variation between emf and logarithmic CO concentration within these limits. It has also been confirmed to be the mixed potential model exhibiting high selectivity for CO among other gases, with quick response and recovery characteristics at 600°C.

In a similar fashion, Park et al. [135] have made use of another semiconducting metal oxide electrode, titania in a very efficient manner so as to obtain excellent selectivity and sensitivity for CO at 500°C. Titania has been used in two forms in the two electrodes: first was the pure form ($TiO_2$) that has been directly used as one electrode. In the other form, titania has been mixed with yttria and Pd to form the other electrode, which has been named as composite titania electrode, TYP. The composition of titania:yttria:Pd was 85:10:5 in weight ratio. The sensing electrodes of titania & TYP have been prepared in a manner so as to acquire high porosity, which proved to be a key feature in enhancing the gas adsorption and hence the ppm level sensitivity. The above combination of electrodes has been used with three different types of current collectors – Pt, Au & $TiO_2$/TYP with the following configurations described in Fig. 40 [135] and their effects on sensing traits were examined at 500°C [135].

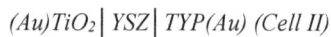

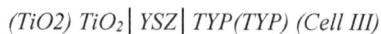

$$(Pt)TiO_2 \mid YSZ \mid TYP(Pt) \ (Cell \ I)$$

$$(Au)TiO_2 \mid YSZ \mid TYP(Au) \ (Cell \ II)$$

$$(TiO2) \ TiO_2 \mid YSZ \mid TYP(TYP) \ (Cell \ III)$$

The voltage response of Cell I containing the catalytic Pt as a current collector was far better than the Cell II that possessed non-catalytic Au contacts. Besides the voltage response, sensing capacity was also higher for Cell I as compared to Cell II, with the former capable of sensing very low CO concentrations (1.0 ppm) up to 1000 ppm without saturation. But Cell II with Au as current collector displayed saturation at high concentrations of CO gas. The sensor devised using novel $TiO_2$ & TYP electrodes along with catalytic Pt contacts exhibited remarkable sensing performance with quick response and good reproducibility [135].

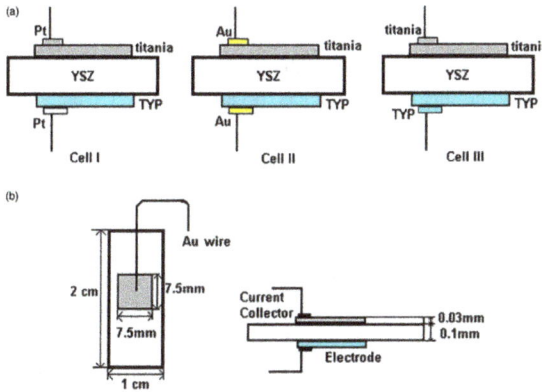

***Fig. 40.*** *Diagrammatic representation of (a) configuration of the cell, (b) structure of three potentiometric solid-state sensors using TiO₂/TYP electrode combination along with Pt, Au & TiO₂/TYP as current collectors. Reproduced here with the permission of John Wiley and Sons [135].*

Among other recent studies, a distinctly innovative idea has been reported by Mahendraprabhu et al. [136] regarding the synthesis of YSZ (Yttria-stabilized zirconia) based potentiometric sensors constituting Pt as reference electrode and ZnO nanospheres (prepared by sol-gel method) as the sensing electrode (Fig. 41) [136].

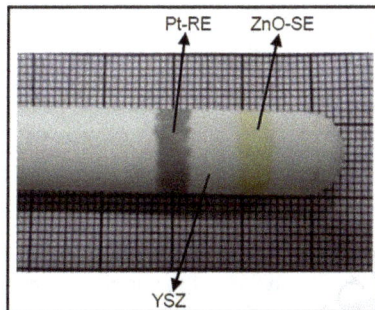

***Fig. 41.*** *Picture of the YSZ based potentiometric sensor comprising of ZnO nanospheres as a sensing electrode. Reproduced here with the permission of Elsevier [136].*

Materials Research Forum LLC
https://doi.org/10.21741/9781644901212

The configuration of the electrochemical cell can be described as:

5 vol% $O_2$, ZnO │ YSZ │ Pt, 5 vol% $O_2$ (in base gas)

CO + 5 vol% $O_2$, ZnO │ YSZ │ Pt, 5 vol% $O_2$ + CO        (in the sample gas)

The gas response ($\Delta V_s$) is interpreted as the difference between the electric potential of the sensor in CO gas and that in the base gas. It can be expressed as:

$$Gas\ response\ (\Delta V_s) = \Delta V_{CO} - \Delta V_{base\ gas}$$

where $\Delta V_{CO}$ = electric potential of the sensor in CO gas

$\Delta V_{base\ gas}$ = electric potential of the sensor in base gas (5 vol% $O_2$ + $N_2$ balance)

Fig. 42 [136] elucidates the linear variation between the voltage response of the sensor and the logarithmic scale of CO concentration in the range 50-400 ppm at 700°C.

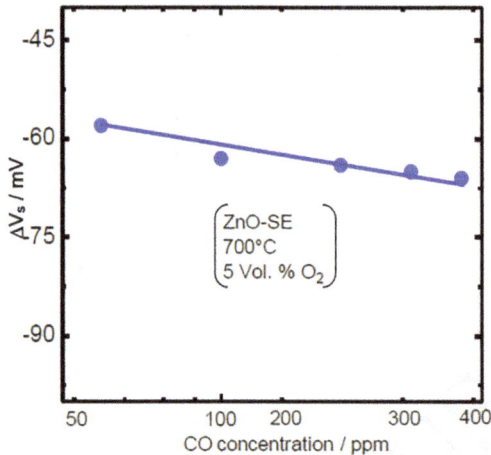

***Fig. 42**. Sensor voltage response relationship with the logarithmic concentration of the analyte CO gas. Reproduced here with the permission of Elsevier [136].*

The direct dependence of sensor voltage on "ln [CO]" is characteristic of a mixed potential type potentiometric sensor. The sensor has been found to be absolutely stable for more than 40 days and produced 90% of the original response with exactly the same rapidity and reproducibility. It has also been demonstrated to possess high selectivity for CO gas as Fig. 43 [136] displays only a partial decomposition in the case of CO in contrast to complete decomposition of other gases (NO, $NO_2$, $C_3H_8$, $C_3H_6$) before reaching the ZnO/YSZ interface(the region responsible for sensing). This can potentially be due to the nano-sizecharacteristic of ZnO.

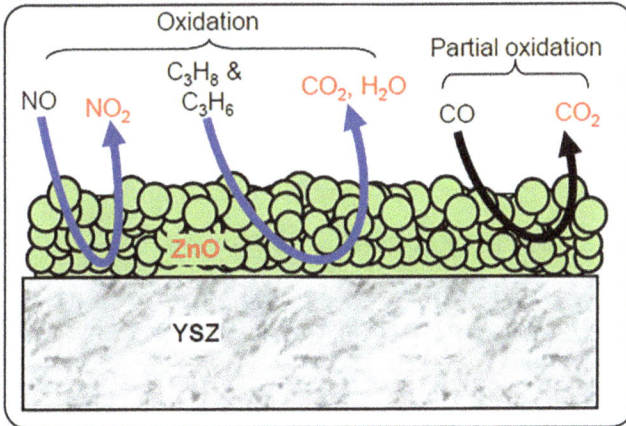

*Fig. 43. Diagram illustrating the selectivity of the sensor towards CO in the presence of other gases. Reproduced here with the permission of Elsevier [136].*

Hence, only CO with no or minimal amounts of interfering gases can reach the sensing surface leading to excellent selectivity of the sensor [136].

In continuation of the ongoing research in this area, Anggraini et al. [137] have tabled a research report on a new YSZ-based planar sensor prepared by them using a couple of metal oxides containing some fractions of Au in their composition as sensing electrodes for detection of CO gas [137]. The cell may be represented as follows:

$$(+) \ Nb_2O_5(+Au)\text{--}SE/YSZ/NiO(+Au')\text{--}SE' \ (-)$$

where $Nb_2O_5(+Au)$–SE was connected to +ve terminal of the electrometer

and $NiO(+Au')$-SE′ was attached to the -ve terminal

Fig. 44 [137] represents the construction of the sensor. The sensor with a combination of two composite electrodes was analyzed for its CO detection properties under humid conditions operated at 450°C and was found to respond to even 10ppm of CO with selectivity (Fig. 45) [137].

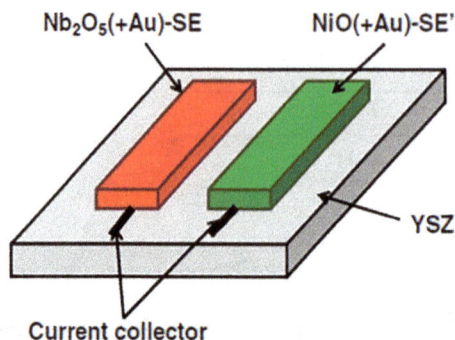

*Fig. 44. YSZ based combined-type sensor with coupled $Nb_2O_5$ (+Au) and NiO (+Au')
electrodes. Reproduced here with the permission of Elsevier [137].*

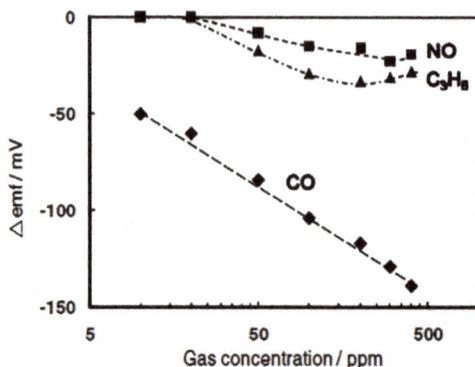

*Fig. 45. Graphical variation of the YSZ based combined type sensor voltage response
with the concentration of CO gas in the presence of other gases at 450 °C operating
temperature under humid conditions. Reproduced here with the permission of Elsevier
[137].*

All the sensors discussed in this category till now have been based on a mixed potential
type mechanism utilizing the stabilized zirconia as the solid-state electrolyte. In the same
category, Hyodo et al. [138] have investigated a variation in the solid electrolyte. An
anion conducting polymer has been used as an electrolyte during the study. Various metal
oxides ($CeO_2$, $Bi_2O_3$, $In_2O_3$, $SnO_2$, $V_2O_5$) have been investigated for sensing electrode-

based material along with an examination of the effect of noble metal loading on the metal oxide electrode. The elementary design of the sensor is illustrated in Fig. 46 [138].

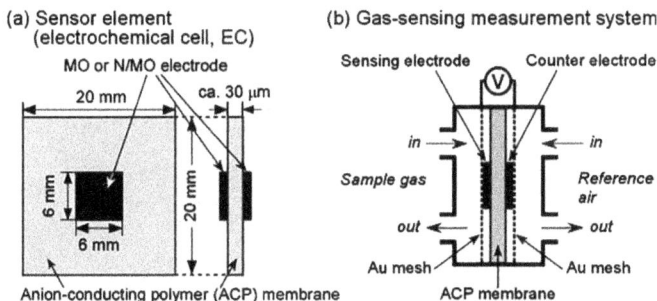

***Fig. 46****. Pictorial representation of (a) anion conducting polymer (ACP) as a sensing element, (b) overall design of the gas sensor. Reproduced here with the permission of Elsevier [138].*

The detailed study of multiple oxides with a range of loading metals resulted in some inferences. The $CeO_2$ & $Bi_2O_3$ electrodes were found to be largely sensitive to CO, but they did not show any change in sensitivity upon the loading of inert metals. On the other hand, high selectivity towards CO has been observed only in $Pt/SnO_2$ sensor [138].

**(b) Amperometric sensors**

Amperometric sensors generally contain a set of three electrodes viz. a working electrode, a reference & a counter electrode which are all connected through an electrolyte [139]. In some simpler configurations, reference and counter electrodes can be combined to form a single electrode. The electrodes utilized in the amperometric gas sensors are of two types:

- Gas diffusion electrode: It is defined as a porous electrode that allows analyte gas diffusion to the reaction site where the electrode is in contact with electrolyte. It is prepared using polytetrafluoroethylene (PTFE) powder mixed with finely divided metals.
- Solid polymer electrolyte-electrode (SPE-electrode): It is defined as a porous layer of metal attached to a solid polymer electrolyte (ion-conducting membrane). The electrode metal layer is composed of Au or Pt which is prepared by a process of reduction of a suitable salt. The electrodes are set up in the cell such that it contacts the gas phase from the front side and liquid electrolyte from the backside [139].

Fig. 47 [140] depicts an example of an amperometric gas sensor for CO composed of a single porous electrode.

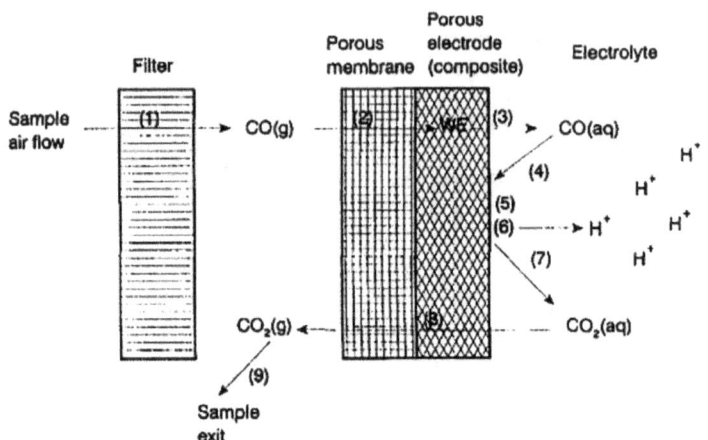

**Fig. 47.** *Amperometric CO gas sensor made of a single porous electrode as a working electrode. Reproduced here with the permission of John Wiley and Sons [140].*

The electrolyte in the amperometric gas sensor can be liquid, solid or gel [141]. Nafion is an example of a solid polymer electrolyte that has been extensively employed in these sensors as a membrane or film. It is referred to as a copolymer of PTFE and polysulphonylfluoride vinyl ether. The sulphonic acid groups attached to the ether part are chemically joined to the perfluorocarbon backbone. It resembles Teflon in inertness [140]. Amperometric detection principle has utilized by Otagawa et al. [142] to fabricate a series of miniaturized sensors consisting of three Pt electrodes and solution cast-nafion film as solid polymer electrolyte. Four variants of substrates (smooth glass, smooth ceramic, rough ceramic and sandblasted glass) and three varieties of solution cast-nafion films were examined for the CO micro-electrochemical sensor with five different sizes of gap (5-50µm) between the adjacent microelectrodes. The fundamental structure of the planar sensor is described in Fig. 48 [142].

Materials Research Forum LLC
https://doi.org/10.21741/9781644901212

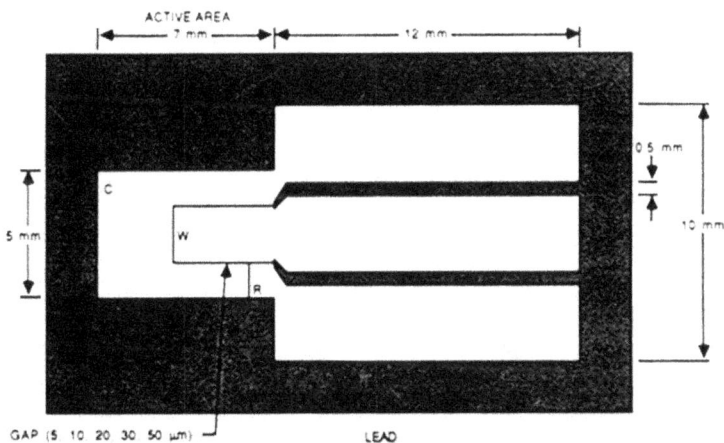

***Fig. 48***. *Fundamental design of the planar microelectrochemical sensor; W= Working/ Sensing electrode, C= Counter electrode, R= Reference electrode. Reproduced here with the permission of Elsevier [142].*

In nafion films, $Na^+$ ions were exchanged by $H^+$ via their acidification before the sensing experiments, in order to facilitate reactions at the electrodes (Scheme 4) [142].

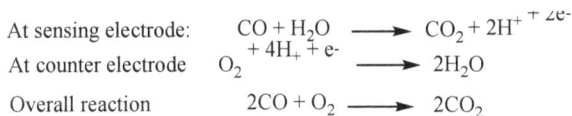

At sensing electrode: $\quad CO + H_2O \longrightarrow CO_2 + 2H^+ + 2e^-$

At counter electrode $\quad O_2 + 4H_+ + e^- \longrightarrow 2H_2O$

Overall reaction $\quad\quad\quad 2CO + O_2 \longrightarrow 2CO_2$

*Scheme 4. The reactions involved during the sensing process [142].*

Out of all the combinations studied, the sensor with a smooth ceramic substrate having a 10μm gap between the electrodes and 1.0 μm thick nafion film (5 wt% Nafion in 95 wt% propanol) gave the finest and reproducible sensing results (Fig. 49)[142].

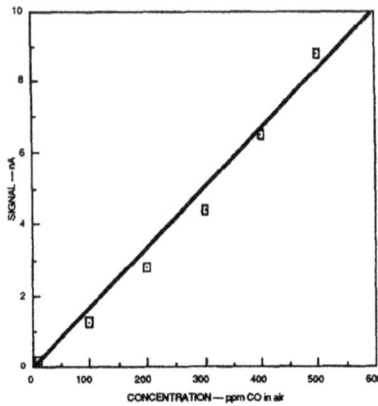

***Fig. 49****. Sensor response dependence on the concentration of CO at -50 mV bias and 45% relative humidity. Reproduced here with the permission of Elsevier [142].*

The sensitivity of this variant was of the order of 8.0 pA/ppm with 70s response time, which are not satisfactory. Another problem was related to the stability of the sensor with time, it did not give steady response characteristics on a long-term basis [142]. This drawback has overcome by Heqing et al. [143] who presented an equivalent work on fabrication SPE-based amperometric sensor using the nafion membrane as an electrolyte and a hydrophilic teflon based Pt black electrode as the sensing electrode. Fig. 50 [143] describes the sensor's current response to CO concentration in different ranges.

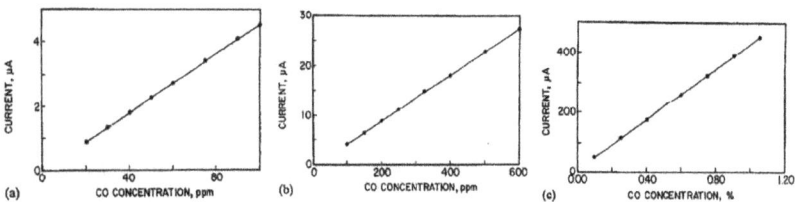

***Fig. 50****. Sensor current response to CO in concentration ranges (a) 20-100ppm, (b) 100-500ppm, (c) 0.1-1.04% CO. Reproduced here with the permission of Elsevier [143].*

The response time of the sensor has been found to be rather quick with a value of 3s. Fig. 51[143]illustrates the stability of the sensor with time.

According to the graph, the current signal of the sensor decreased by 15% over 40 days span and then remained stable for the next 50 days i.e., a fairly stable response for 90 days [143].

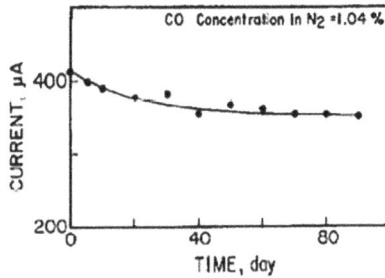

*Fig. 51. Long term stability characteristics of the sensor. Reproduced here with the permission of Elsevier [143].*

Better innovation has been reported by Vander Wal et al. [144]. The investigators have also designed a nafion-SPE based CO sensor and tested the performance of both forms of nafion – the dry and the wet polymer [144]. It has been noticed that the dry nafion sensor was not as stable as the wet one, which exhibited immense stability over a long period of time. Platinum black electrodes were exercised in the sensor. Fig. 52 [144] describes the layout of the wet nafion based sensor.

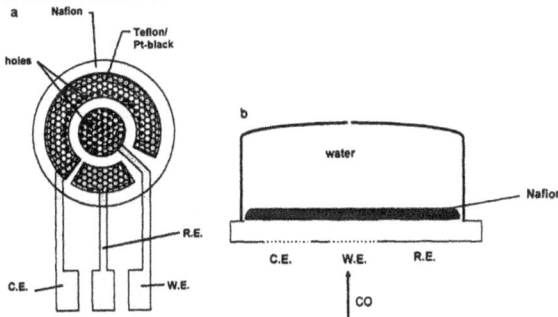

*Fig. 52. (a) View of the sensor from top displaying the holes, Nafion, Teflon bound Pt black electrodes and Pt current collectors, (b) cross-sectional view presenting the ceramic substrate with holes, Nafion layer, electrodes and the water reservoir. Reproduced here with the permission of Elsevier [144].*

The wet nafion sensor was found to show sensitivity to sub-ppm levels of CO. The sensor response was also checked for long term stability, which revealed uniform signal characteristics for over a year, displaying exceptional stability (Fig. 53) [144].

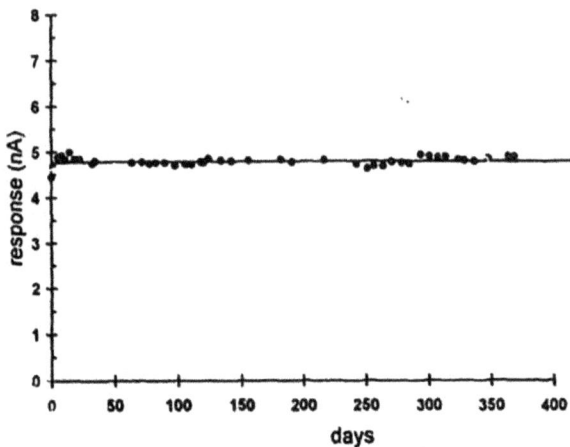

**Fig. 53**. *The study of sensor performance with time. The sensitivity has been calculated on the basis of ppm of carbon monoxide. Reproduced here with the permission of Elsevier [144].*

Santhosh et al. [145] have brought a completely novel amperometric sensor based on multiwalled carbon nanotubes grafted with polydiphenylamine modified electrode (MWNT-g-PDPA-ME), which performed excellently in sensing CO gas. MWNTs have been appropriately functionalized and grafted with polydiphenylamine (PDPA) chains on glassy carbon (GC) electrode by the process of electrochemical polymerization technique. Fig. 54 [145] schematically represents the process of MWNT-g-PDPA-ME fabrication via a two-step process. The first step involved the amine functionalization of MWNT to form MWNT-NH$_2$. The second step utilized the cyclic voltammetry technique for the deposition of MWNT-g-PDPA film on the surface of the GC electrode to get MWNT-g-PDPA-ME as the working electrode [145].

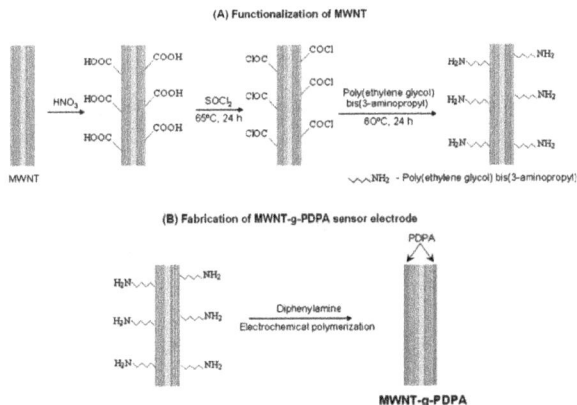

***Fig. 54****. Scheme of MWNT-g-PDPA-ME synthesis. Reproduced here with the permission of Elsevier [145].*

The synthesized electrode has been characterized using cyclic voltammetry, impedance spectroscopy, and chronoamperometry. The sensor contained the electrochemical cell in which MWNT-g-PDPA-ME has been made the working electrode, SCE - the reference and Pt wire – the counter electrode. The amperometric sensing measurements of the sensor electrode towards CO have been carried out in 0.5M HClO$_4$. The limit of detection has been calculated to be 0.01ppm with very rapid response and recovery signals. The sensor initially encountered significant interferences from gases like CH$_4$, NH$_3$, C$_3$H$_8$, N$_2$O (500 ppm), which have been minimized to a negligible level by using 0.5% Nafion coated MWNT-g-PDPA-ME, making it highly selective for CO. Moreover, the sensor response was also found to be stable for a sufficiently long period [145].

A further distinct approach has been introduced by Chou et al. [146] who constructed an amperometric CO sensor by deposition of Pt nanoparticles on SPUME (Screen printed edge band ultra microelectrode) and utilized nafion as the solid polymer electrolyte (Fig. 55) [146].

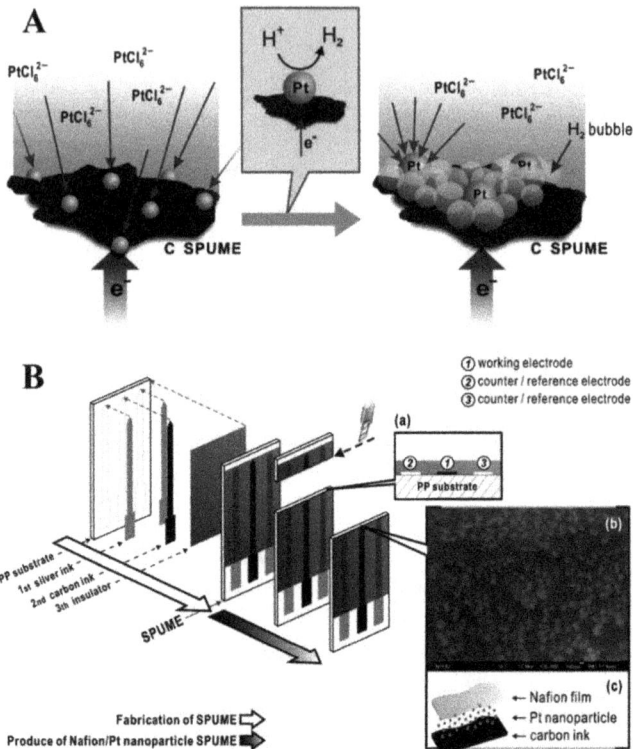

*Fig. 55.* *(A) Demonstration of the process of electrodeposition of Pt nanoparticles on SPUME, (B) (a) SPUME assembly comprising of in-built 3 electrode configuration, (b) SEM photograph of Pt deposited on Carbon WE surface, (c) cartoon representation of the electrode. Reproduced here with the permission of John Wiley and Sons [146].*

The amperometric response of the sensor has been observed to be linear with respect to the concentration of CO up to 1000ppm. The sensitivity of the NPt-SPUME sensor has been reported to be 3.76 nA/ppm/cm$^2$ and the limit of detection has been observed to be 28.6ppm.

**(vii) Photoacoustic gas sensors**

Photoacoustic spectroscopy (PAS) is a technique in which electromagnetic radiation impinges on the matter (gas, liquid or solid) and its absorption by the matter results in the generation of an acoustic signal (acoustic pressure wave) [147]. The pressure wave is produced when the sample (e.g. gas) absorbs the radiation, gets heated up and consequently expands [148].

Quartz enhanced PAS (QEPAS) is a variant of PAS in which a quartz tuning fork is employed as a transducer. When laser radiation is directed between the tines of the fork, the analyte gas absorbs it resulting in a recurrent thermal expansion, which generates a weak acoustic pressure wave. This pressure wave is converted to an electrical signal by the tuning fork via the piezoelectric effect [149]. Its magnitude directly corresponded to the concentration of the analyte gas, which can be analyzed by the detector [148].

QEPAS based sensors are deployed for the detection of gases for monitoring the quality of air [149]. Ma et al. [150] have demonstrated QEPAS based CO detection by manufacturing a compact and sensitive sensor utilizing a mid-IR all-fiber structure. The benefits of using an all-fiber composition include easier spatial alignment, lessor loss in insertion, improved system stability, compressed sensor size and feasible cost.In addition to the above modification, the use of a 3D printed acoustic module resulted in reduced sensor volume (3.5 cm$^3$) and 5.0 g weight.

Carbon monoxide (CO) detection involves the basis that it undergoes 3 different vibrational transitions, namely as a fundamental band (υ) near 4.6μm [151], first overtone (2υ) around 2.33μm [152] and second overtone (3υ) around 1.57μm [153]. The fundamental vibrational transition requires cost-intensive quantum cascade lasers (QCLs), while the first and second overtones require lesser expensive diode lasers. In this sensor, a 2.33μmdistributed feedback (DFB) fiber-coupled diode laser has been used as the excitation source. The transducer used was a Quartz tuning fork (QTF) with a 200μm gap. An acoustic micro-resonator was also employed for acoustic wave enhancement. The rate of vibrational-translational relaxation of the analyte(CO) molecule was accelerated by the use of water vapours.

From the graph, it can be inferred that the injection of water vapour concentration of 1.53% yielded maximum enhancement (11-fold) in the signal response. The all-fiber QEPAS-CO sensor can display a minimum detection limit (MDL) of 4.2ppm with 1s integration time. The limit was improved to 1.3 ppm with 150 s integration time [150]. The same group of investigators have struggled to achieve ppb level detection in their other QEPAS based portable, compact and rechargeable devices [154].

Li et al. [155] have come up with a modification in the quartz tuning fork (QTF) in QEPAS based sensor. A new 1.5kHz grooved QTF (Fig. 56) [155] has been exploited in orderto achieve a high-quality factor at atmospheric pressure along with high sensitivity.

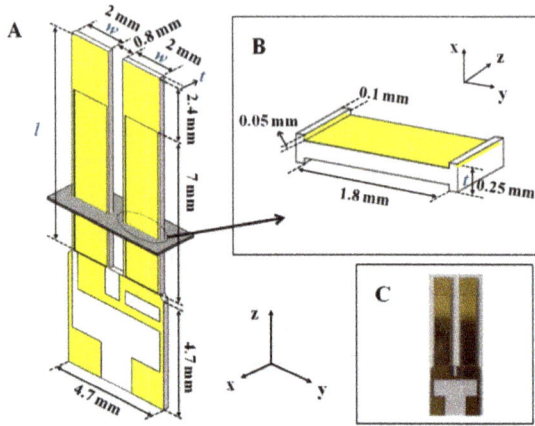

**Fig. 56**. *(A) Dimensions of the grooved QTF. Yellow regions represent the electrode layout section, (B) cross-section of grooves on QTF, (C) Picture of the new grooved QTF. Reproduced here with the permission of ACS [155].*

The QTF of 800μm gap has been selected along with a quantum cascade laser (QCL) as the excitation source. Fig.59 [155] pictorially describes the working of QEPAS based sensor employing a grooved QTF.

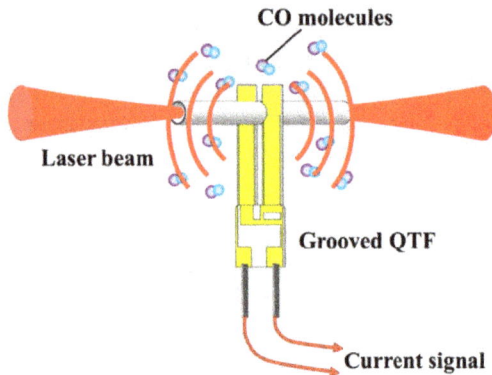

**Fig. 57**. *Pictorial representation of the QEPAS based CO sensor comprising of a grooved QTF. Reproduced here with the permission of ACS [155].*

The detailed block diagram of the QEPAS based sensor system has been expressed in Fig. 58 [155].

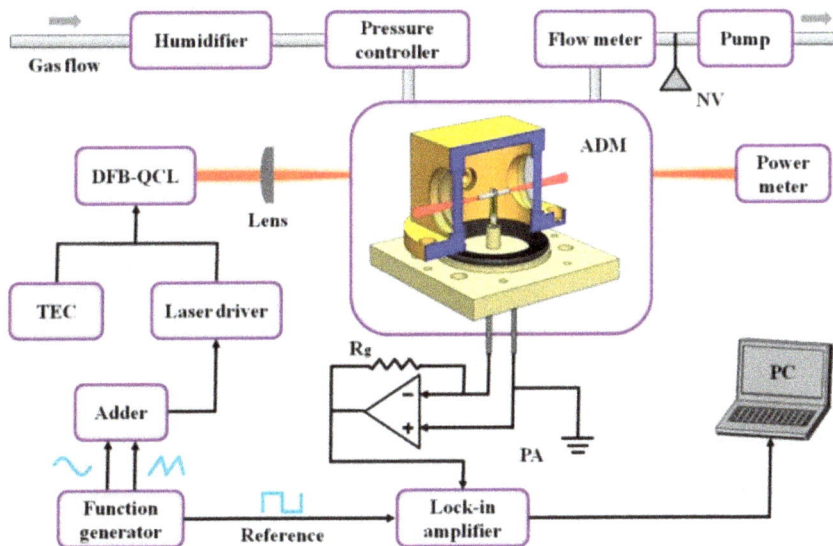

**Fig. 58**. *Block diagram of QEPAS based sensor for CO involving a new grooved QTF; TEC: Temperature Controller, NV: Needle Valve, PA: Preamplifier. Reproduced here with the permission of ACS [155].*

The sensor has been evaluated thoroughly to conclude that it was capable of detecting 7.0 ppb CO for 300ms average time with the addition of 2.5% water vapour. It was also known to exhibit rapid response and adequate stability. The carbon monoxide concentration has been monitored continuously for a period of seven days using the NDIR method (with 50 ppb detection limit, Fig. 59A [155]) and compared with the method used by China National Environmental Monitoring Center (CNEMC) (Fig. 59B [155]). An excellent agreement between Fig. 59A and Fig. 59B indicated the response of the sensor towards carbon dioxide monitoring.

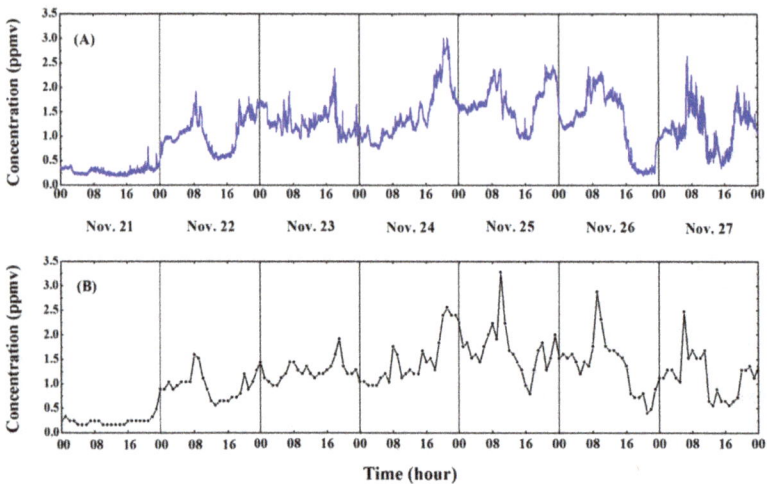

**Fig. 59**. *(A) Monitoring of carbon monoxide concentration for a weak using QEPAS based sensor at Shanxi University campus. (B) The data recorded by the Department of Ecology and Environment of Shanxi Province during the same period. Reproduced here with the permission of ACS [155].*

## 3. Biosensing of carbon monoxide

A typical biosensor usually has a biological recognition component along with a physical transducer to translate a biological response into an optical, thermal or an electrical signal, which can be used to quantify the concentration of an analyte [156-158]. Based on the recognition process, the biosensors are characterized as enzymatic biosensors, immunosensors, biomimetic biosensors, cell-based biosensors, and DNA biosensors, while based on transducer component, biosensors are further classified as optical, thermal, piezoelectric and electrochemical biosensors [156, 159-163].

Heme proteins have been found to serve as an important tool in molecular biosensors [164-166]. Michael Chan [165] has summarised the invention of heme-proteins as potential molecular sensors that can be utilized in biosensing applications. Shelver et al. [167] have investigated heme protein namely CooA of Rhodospirillum rubrum (a phototrophic bacterium which gains energy from CO for its anaerobic growth)for the recognition of carbon monoxide. CooA belongs to the family of transcriptional regulators, which are similar to cAMP receptor protein and fumavate nitrate reduction proteins in *E. Coli.* The mechanism of working of the biosensor exploited the difference in conformation of the heme protein before and after the analyte binding. The heme-analyte binding served as the input signal and the resulting conformational change in the heme structure responsible for activation of various enzymatic activities as the output [165, 167]. The study involves the wild-type CooA purification up to 95% purity from R. rubrum (native organism). The purified protein interacts with carbon monoxide through heme, which helps it to bind the sequence-specific DNA. However, in the absence of carbon monoxide, the binding of protein with DNA fails. The interaction between the heme of CooA and carbon monoxide causes a shift in the absorption spectrum. The pH dependence for the binding was investigated between pH 4-10 range, which indicated 4 to be the optimum pH [168].

## Conclusions

Keeping in consideration the life-threatening consequences of the highly poisonous carbon monoxide gas, sensors began to be developed for its recognition. A large variety of technologies have been exploited to fabricate these sensing devices till today, namely metal oxide semi conduction, conducting properties of carbon nanotubes & polymeric thin films, colorimetry, potentiometry and amperometry among others. Non-dispersive infrared spectroscopy has also been probed for sensing applications. Vast innovations and discoveries regarding their continuing progress are being presented every new day. Most of these are discussed in this chapter. Photoacoustic spectroscopy has emerged as a promising technique for the development of highly selective and sensitive sensors for carbon monoxide. This specific area has not witnessed much research in the past, so it possesses great potential for the development of sensors with utmost sensitivity and selectivity in the future.

## Acknowledgments

The authors sincerely thank DST-SERB and DRDO, India for financial support (EMR/2016/005022 and ERIP/ER/DG-NSM/990116702/M/01/1645). The authors are thankful to the Principal, St. Stephen's College for the necessary infrastructure.

## References

[1] L.D. Prockop, R.I. Chichkova, Carbon monoxide intoxication: An updated review, J. Neurol. Sci. 262 (2007) 122-130. https://doi.org/10.1016/j.jns.2007.06.037

[2] A.L. Chiew, N.A. Buckley, Carbon monoxide poisoning in the 21st century, Crit. Care 18 (2014) 221-221. https://doi.org/10.1186/cc13846

[3] A. Ernst, J.D. Zibrak, Carbon monoxide poisoning, N. Engl. J. Med. 339 (1998) 1603-1608. https://doi.org/10.1056/NEJM199811263392206

[4] C. Pulce, J. Descotes, Carbon monoxide: The unnoticed poison of the 21st century, J. Environ. Med. 1 (1999) 57-57. https://doi.org/10.1002/(SICI)1099-1301(199901/03)1:1<57::AID-JEM2>3.0.CO;2-L

[5] A. Sobhakumari, R.H. Poppenga, J.B. Pesavento, F.A. Uzal, Pathology of carbon monoxide poisoning in two cats, BMC Vet. Res. 14 (2018) 67-67. https://doi.org/10.1186/s12917-018-1385-4

[6] D.R. Varma, S. Mulay, S. Chemtob, Carbon Monoxide: From public health risk to painless killer, in: R.C. Gupta (Ed.), Handbook of toxicology of chemical warfare Agents (Second Edition), Academic Press, Boston, 2015, pp. 267-286

[7] J.J. Rose, L. Wong, Q. Xu, C.F. Mctiernan, S. Shiva, J. Tejero, M.T. Gladwin, Carbon monoxide poisoning: pathogenesis, management, and future directions of therapy, Am. J. Respir. Crit. Care Med. 195 (2017) 596-606.

[8] J.A. Donald, Subchapter,Carbon monoxide, in: Y. Takei, H. Ando, K. Tsutsui (Eds.), Handbook of hormones, Academic Press, San Diego, 2016, pp. 606-e103B-3

[9] D. Silverstein, K. Hopper, Small animal critical care medicine-E-Book, Elsevier Health Sciences, St. Louis, Missouri,2014

[10] A. El-Hellani, S. Al-Moussawi, R. El-Hage, S. Talih, R. Salman, A. Shihadeh, N.A. Saliba, Carbon monoxide and small hydrocarbon emissions from sub-ohm electronic cigarettes,Chem. Res. Toxicol. 32 (2019) 312-317. https://doi.org/10.1021/acs.chemrestox.8b00324

[11] A.L. Jones, P.I. Dargan, Poisoning: Overview and statistics, in: J. Payne-James, R.W. Byard (Eds.), Encyclopedia of forensic and legal medicine (Second edition), Elsevier, Oxford, 2016, pp. 1-9

[12] P.A. Rodgers, H.J. Vreman, P.A. Dennery, D.K. Stevenson, Sources of carbon monoxide (CO) in biological systems and applications of CO detection technologies, Semin. Perinatol, 1994, pp. 2-10.

[13] J.G. Olivier, J.P.J. Bloos, J.J. Berdowski, A.J. Visschedijk, A.F. Bouwman, A 1990 global emission inventory of anthropogenic sources of carbon monoxide on 1× 1 developed in the framework of EDGAR/GEIA, Chemosphere Global Change SCI. 1 (1999) 1-17. https://doi.org/10.1016/S1465-9972(99)00019-7

[14] D.G. Penney, Carbon monoxide poisoning, CRC Press, Boca Raton, Florida, 2019

[15] S.W. Salyer, Chapter 17 - Toxicology emergencies, in: S.W. Salyer (Ed.), Essential Emergency Medicine, W.B. Saunders, Philadelphia, 2007, pp. 923-1049

[16] J.L. Cadet, K.I. Bolla, Environmental toxins and disorders of the nervous system, in: A.H.V. Schapira, E. Byrne, S. DiMauro, R.S.J. Frackowiak, R.T. Johnson, Y. Mizuno, M.A. Samuels, S.D. Silberstein, Z.K. Wszolek (Eds.), Neurology and clinical neuroscience, Mosby, Philadelphia, 2007, pp. 1477-1488

[17] K.I. Bolla, J.L. Cadet, Chapter 39 - Exogenous acquired metabolic disorders of the nervous system: Toxins and illicit drugs, in: C.G. Goetz (Ed.), Textbook of clinical neurology (Third edition), W.B. Saunders, Philadelphia, 2007, pp. 865-896

[18] L. Rahilly, D.C. Mandell, Carbon monoxide, Small animal critical care medicine, Elsevier, St. Louis, Missouri,2009, pp. 369-373

[19] L. Prockop, Carbon Monoxide, in: M.R. Dobbs (Ed.), Clinical Neurotoxicology, W.B. Saunders, Philadelphia, 2009, pp. 500-514

[20] K.T. Fitzgerald, Carbon monoxide, Small animal toxicology, Elsevier, St. Louis, Missouri,2013, pp. 479-487

[21] J.M. Miyasaki, Chapter 26 - Chorea caused by toxins, in: W.J. Weiner, E. Tolosa (Eds.), Handbook of clinical neurology, Elsevier, Amsterdam, Netherlands, 2011, pp. 335-346

[22] M. Hjiri, L. El Mir, S.G. Leonardi, A. Pistone, L. Mavilia, G. Neri, Al-doped ZnO for highly sensitive CO gas sensors, Sens. Actuators B Chem. 196 (2014) 413-420. https://doi.org/10.1016/j.snb.2014.01.068

[23] N. Singh, C. Yan, P.S. Lee, Room temperature CO gas sensing using Zn-doped $In_2O_3$ single nanowire field effect transistors, Sens. Actuators B Chem. 150 (2010) 19-24. https://doi.org/10.1016/j.snb.2010.07.051

[24] T. Seiyama, S. Kagawa, Study on a detector for gaseous components using semiconductive thin films, Anal. Chem. 38 (1966) 1069-1073. https://doi.org/10.1021/ac60240a031

[25] G. Neri, Non-conventional sol-gel routes to nanosized metal oxides for gas sensing: From materials to applications, Sci. Adv. Mater. 2 (2010) 3-15. https://doi.org/10.1166/sam.2010.1062

[26] D.H. Yoon, G.M. Choi, Microstructure and CO gas sensing properties of porous ZnO produced by starch addition, Sens. Actuators B Chem. 45 (1997) 251-257. https://doi.org/10.1016/S0925-4005(97)00316-X

[27] M.C. Horrillo, A. Serventi, D. Rickerby, J. Gutiérrez, Influence of tin oxide microstructure on the sensitivity to reductor gases, Sens. Actuators B Chem. 58 (1999) 474-477. https://doi.org/10.1016/S0925-4005(99)00106-9

[28] C.A. Papadopoulos, D.S. Vlachos, J.N. Avaritsiotis, Effect of surface catalysts on the long-term performance of reactively sputtered tin and indium oxide gas sensors, Sens. Actuators B Chem. 42 (1997) 95-101. https://doi.org/10.1016/S0925-4005(97)00190-1

[29] G. Sberveglieri, G. Faglia, S. Groppelli, P. Nelli, Methods for the preparation of NO, $NO_2$ and $H_2$ sensors based on tin oxide thin films, grown by means of the r.f. magnetron sputtering technique, Sens. Actuators B Chem. 8 (1992) 79-88. https://doi.org/10.1016/0925-4005(92)85012-L

[30] S.-W. Choi, S.S. Kim, Room temperature CO sensing of selectively grown networked ZnO nanowires by Pd nanodot functionalization, Sens. Actuators B Chem. 168 (2012) 8-13. https://doi.org/10.1016/j.snb.2011.12.100

[31] D. Kohl, The role of noble metals in the chemistry of solid-state gas sensors, Sens. Actuators B Chem. 1 (1990) 158-165. https://doi.org/10.1016/0925-4005(90)80193-4

[32] G.T. Rao, D.T. Rao, Gas sensitivity of ZnO based thick film sensor to $NH_3$ at room temperature, Sens. Actuators B Chem. 55 (1999) 166-169. https://doi.org/10.1016/S0925-4005(99)00049-0

[33] S.C. Navale, V. Ravi, I.S. Mulla, S.W. Gosavi, S.K. Kulkarni, Low temperature synthesis and NOx sensing properties of nanostructured Al-doped ZnO, Sens. Actuators B Chem. 126 (2007) 382-386. https://doi.org/10.1016/j.snb.2007.03.019

[34] J.F. Chang, H.H. Kuo, I.C. Leu, M.H. Hon, The effects of thickness and operation temperature on ZnO:Al thin film CO gas sensor, Sens. Actuators B Chem. 84 (2002) 258-264. https://doi.org/10.1016/S0925-4005(02)00034-5

[35] Z. Yang, Y. Huang, G. Chen, Z. Guo, S. Cheng, S. Huang, Ethanol gas sensor based on Al-doped ZnO nanomaterial with many gas diffusing channels, Sens. Actuators B Chem. 140 (2009) 549-556. https://doi.org/10.1016/j.snb.2009.04.052

[36] P.P. Tsai, I.C. Chen, M.H. Tzeng, Tin oxide (SnOX) carbon monoxide sensor fabricated by thick-film methods, Sens. Actuators B Chem. 25 (1995) 537-539. https://doi.org/10.1016/0925-4005(95)85116-X

[37] N. Yamazoe, New approaches for improving semiconductor gas sensors, Sens. Actuators B Chem. 5 (1991) 7-19. https://doi.org/10.1016/0925-4005(91)80213-4

Materials Research Forum LLC
https://doi.org/10.21741/9781644901212

[38] C.S. Prajapati, D. Visser, S. Anand, N. Bhat, Honeycomb type ZnO nanostructures for sensitive and selective CO detection, Sens. Actuators B Chem. 252 (2017) 764-772. https://doi.org/10.1016/j.snb.2017.06.070

[39] J. Zhang, X. Liu, G. Neri, N. Pinna, Nanostructured materials for room-temperature gas sensors, Adv. Mater. 28 (2016) 795-831. https://doi.org/10.1002/adma.201503825

[40] S.A. Akbar, L.B. Younkman, Sensing mechanism of a carbon monoxide sensor based on anatase titania, J. Electrochem. Soc. 144 (1997) 1750-1753. https://doi.org/10.1149/1.1837673

[41] L.D. Birkefeld, A.M. Azad, S.A. Akbar, Carbon monoxide and hydrogen detection by anatase modification of titanium dioxide, J. Am. Ceram. Soc. 75 (1992) 2964-2968. https://doi.org/10.1111/j.1151-2916.1992.tb04372.x

[42] N.O. Savage, S.A. Akbar, P.K. Dutta, Titanium dioxide based high temperature carbon monoxide selective sensor, Sens. Actuators B Chem. 72 (2001) 239-248. https://doi.org/10.1016/S0925-4005(00)00676-6

[43] T. Anukunprasert, C. Saiwan, E. Traversa, The development of gas sensor for carbon monoxide monitoring using nanostructure of Nb–TiO2, Sci. Technol. Adv. Mat. 6 (2005) 359-363. https://doi.org/10.1016/j.stam.2005.02.020

[44] A. Salehi, A highly sensitive self heated $SnO_2$ carbon monoxide sensor, Sens. Actuators B Chem. 96 (2003) 88-93. https://doi.org/10.1016/S0925-4005(03)00490-8

[45] S. Peng, P. Hong, Y. Li, X. Xing, Y. Yang, Z. Wang, T. Zou, Y. Wang, Pt decorated $SnO_2$ nanoparticles for high response CO gas sensor under the low operating temperature, J. Mater. Sci.: Mater. Electron. 30 (2019) 3921-3932. https://doi.org/10.1007/s10854-019-00677-7

[46] A. Debataraja, N.L.W. Septiani, B. Yuliarto, Nugraha, B. Sunendar, H. Abdullah, High performance of a carbon monoxide sensor based on a Pd-doped graphene-tin oxide nanostructure composite, Ionics 25(2019) 4459-4468. https://doi.org/10.1007/s11581-019-02967-w

[47] L. Cheng, M.-W. Shao, D. Chen, D.D. Duo Ma, S.-T. Lee, $SnO_2$ nanowires with strong yellow emission and their application in photoswitches, CrystEngComm 12 (2010) 1536-1539. https://doi.org/10.1039/B911664H

[48] B. Kim, Y. Lu, A. Hannon, M. Meyyappan, J. Li, Low temperature $Pd/SnO_2$ sensor for carbon monoxide detection, Sens. Actuators B Chem. 177 (2013) 770-775.

https://doi.org/10.1016/j.snb.2012.11.020

[49] W. Zeng, Y. Li, B. Miao, L. Lin, Z. Wang, Recognition of carbon monoxide with $SnO_2$/Ti thick-film sensor and its gas-sensing mechanism, Sens. Actuators B Chem. 191 (2014) 1-8. https://doi.org/10.1016/j.snb.2013.09.092

[50] K. Wang, T. Zhao, G. Lian, Q. Yu, C. Luan, Q. Wang, D. Cui, Room temperature CO sensor fabricated from Pt-loaded $SnO_2$ porous nanosolid, Sens. Actuators B Chem. 184 (2013) 33-39. https://doi.org/10.1016/j.snb.2013.04.054

[51] C.T. Wang, M.T. Chen, Vanadium-promoted tin oxide semiconductor carbon monoxide gas sensors, Sens. Actuators B Chem. 150 (2010) 360-366. https://doi.org/10.1016/j.snb.2010.06.060

[52] R.K. Sharma, P.C.H. Chan, Z. Tang, G. Yan, I.M. Hsing, J.K.O. Sin, Sensitive, selective and stable tin dioxide thin-films for carbon monoxide and hydrogen sensing in integrated gas sensor array applications, Sens. Actuators B Chem. 72 (2001) 160-166. https://doi.org/10.1016/S0925-4005(00)00646-8

[53] S. Bose, S. Chakraborty, B. Ghosh, D. Das, A. Sen, H.S. Maiti, Methane sensitivity of Fe-doped $SnO_2$ thick films, Sens. Actuators B Chem. 105 (2005) 346-350. https://doi.org/10.1016/j.snb.2004.06.023

[54] J.H. Yu, G.M. Choi, Selective CO gas detection of CuO-and ZnO-doped $SnO_2$ gas sensor, Sens. Actuators B Chem. 75 (2001) 56-61. https://doi.org/10.1016/S0925-4005(00)00742-5

[55] Z. Ansari, S. Ansari, T. Ko, J.-H. Oh, Effect of $MoO_3$ doping and grain size on $SnO_2$-enhancement of sensitivity and selectivity for CO and $H_2$ gas sensing, Sens. Actuators B Chem. 87 (2002) 105-114. https://doi.org/10.1016/S0925-4005(02)00226-5

[56] P.G. Harrison, M.J. Willett, The mechanism of operation of tin(iv) oxide carbon monoxide sensors, Nature 332 (1988) 337-339. https://doi.org/10.1038/332337a0

[57] W. Göpel, K.D. Schierbaum, $SnO_2$ sensors: current status and future prospects, Sens. Actuators B Chem. 26 (1995) 1-12. https://doi.org/10.1016/0925-4005(94)01546-T

[58] F. Pourfayaz, A. Khodadadi, Y. Mortazavi, S.S. Mohajerzadeh, $CeO_2$ doped $SnO_2$ sensor selective to ethanol in presence of CO, LPG and $CH_4$, Sens. Actuators B Chem. 108 (2005) 172-176. https://doi.org/10.1016/j.snb.2004.12.107

[59] A. Khodadadi, S.S. Mohajerzadeh, Y. Mortazavi, A.M. Miri, Cerium oxide/$SnO_2$-based semiconductor gas sensors with improved sensitivity to CO, Sens. Actuators

B Chem. 80 (2001) 267-271. https://doi.org/10.1016/S0925-4005(01)00915-7

[60] E.S. Putna, R.J. Gorte, J.M. Vohs, G.W. Graham, Evidence for enhanced dissociation of CO on Rh/Ceria, J. Catal. 178 (1998) 598-603. https://doi.org/10.1006/jcat.1998.2206

[61] S.M.A. Durrani, M.F. Al-Kuhaili, I.A. Bakhtiari, M.B. Haider, Investigation of the carbon monoxide gas sensing characteristics of tin oxide mixed cerium oxide thin films, Sensors 12(2012) 2598-2609. https://doi.org/10.3390/s120302598

[62] D. Barreca, A. Gasparotto, C. Maccato, C. Maragno, E. Tondello, E. Comini, G. Sberveglieri, Columnar $CeO_2$ nanostructures for sensor application, Nanotechnology 18(2007) 125502. https://doi.org/10.1088/0957-4484/18/12/125502

[63] S.M.A. Durrani, M.F. Al-Kuhaili, I.A. Bakhtiari, Carbon monoxide gas-sensing properties of electron-beam deposited cerium oxide thin films, Sens. Actuators B Chem. 134 (2008) 934-939. https://doi.org/10.1016/j.snb.2008.06.049

[64] D. Perednis, L.J. Gauckler, Thin film deposition using spray pyrolysis, J. Electroceram. 14 (2005) 103-111. https://doi.org/10.1007/s10832-005-0870-x

[65] J.B. Mooney, S.B. Radding, Spray pyrolysis processing, Annu. Rev. Mater. Sci. 12 (1982) 81-101. https://doi.org/10.1146/annurev.ms.12.080182.000501

[66] A. Tischner, T. Maier, C. Stepper, A. Köck, Ultrathin $SnO_2$ gas sensors fabricated by spray pyrolysis for the detection of humidity and carbon monoxide, Sens. Actuators B Chem. 134 (2008) 796-802. https://doi.org/10.1016/j.snb.2008.06.032

[67] C.T. Wang, H.Y. Chen, Y.C. Chen, Gold/vanadium–tin oxide nanocomposites prepared by co-precipitation method for carbon monoxide gas sensors, Sens. Actuators B Chem. 176 (2013) 945-951. https://doi.org/10.1016/j.snb.2012.10.041

[68] C.T. Wang, J.C. Lin, Surface nature of nanoparticle zinc-titanium oxide aerogel catalysts, Appl. Surf. Sci. 254 (2008) 4500-4507. https://doi.org/10.1016/j.apsusc.2008.01.024

[69] C.-T. Wang, S.-H. Ro, Nanoparticle iron–titanium oxide aerogels, Mater. Chem. Phys. 101 (2007) 41-48. https://doi.org/10.1016/j.matchemphys.2006.02.010

[70] S. Ghosh, M. Narjinary, A. Sen, R. Bandyopadhyay, S. Roy, Fast detection of low concentration carbon monoxide using calcium-loaded tin oxide sensors, Sens. Actuators B Chem. 203 (2014) 490-496. https://doi.org/10.1016/j.snb.2014.06.111

[71] B.K. Min, S.D. Choi, Role of CaO as crystallite growth inhibitor in $SnO_2$, Sens.

Actuators B Chem. 99 (2004) 288-296. https://doi.org/10.1016/j.snb.2003.11.025

[72] M. Yuasa, T. Masaki, T. Kida, K. Shimanoe, N. Yamazoe, Nano-sized PdO loaded $SnO_2$ nanoparticles by reverse micelle method for highly sensitive CO gas sensor, Sens. Actuators B Chem. 136 (2009) 99-104. https://doi.org/10.1016/j.snb.2008.11.022

[73] P. Manjula, S. Arunkumar, S.V. Manorama, Au/$SnO_2$ an excellent material for room temperature carbon monoxide sensing, Sens. Actuators B Chem. 152 (2011) 168-175. https://doi.org/10.1016/j.snb.2010.11.059

[74] C. Xu, J. Tamaki, N. Miura, N. Yamazoe, Grain size effects on gas sensitivity of porous $SnO_2$-based elements, Sens. Actuators B Chem. 3 (1991) 147-155. https://doi.org/10.1016/0925-4005(91)80207-Z

[75] G. Sakai, N. Matsunaga, K. Shimanoe, N. Yamazoe, Theory of gas-diffusion controlled sensitivity for thin film semiconductor gas sensor, Sens. Actuators B Chem. 80 (2001) 125-131. https://doi.org/10.1016/S0925-4005(01)00890-5

[76] N. Matsunaga, G. Sakai, K. Shimanoe, N. Yamazoe, Formulation of gas diffusion dynamics for thin film semiconductor gas sensor based on simple reaction–diffusion equation, Sens. Actuators B Chem. 96(2003) 226-233. https://doi.org/10.1016/S0925-4005(03)00529-X

[77] Y. Bing, Y. Zeng, S. Feng, L. Qiao, Y. Wang, W. Zheng, Multistep assembly of Au-loaded SnO2 hollow multilayered nanosheets for high-performance CO detection, Sens. Actuators B Chem. 227 (2016) 362-372. https://doi.org/10.1016/j.snb.2015.12.065

[78] S. Wang, Y. Wang, J. Jiang, R. Liu, M. Li, Y. Wang, Y. Su, B. Zhu, S. Zhang, W. Huang, A DRIFTS study of low-temperature CO oxidation over Au/$SnO_2$ catalyst prepared by co-precipitation method, Catal. Commun. 10(2009) 640-644. https://doi.org/10.1016/j.catcom.2008.11.009

[79] I. Lisiecki, M.P. Pileni, Synthesis of copper metallic clusters using reverse micelles as microreactors, J. Am. Chem. Soc. 115 (1993) 3887-3896. https://doi.org/10.1021/ja00063a006

[80] K. Osseo-Asare, F. Arriagada, Preparation of $SiO_2$ nanoparticles in a non-ionic reverse micellar system, Colloids Surf. 50 (1990) 321-339. https://doi.org/10.1016/0166-6622(90)80273-7

[81] T. Miyao, N. Toyoizumi, S. Okuda, Y. Imai, K. Tajima, S. Naito, Preparation of Pt/$SiO_2$ ultra-fine particles in reversed micelles and their catalytic activity, Chem.

Lett. 28 (1999) 1125-1126. https://doi.org/10.1246/cl.1999.1125

[82] T. Kida, G. Guan, Y. Minami, T. Ma, A. Yoshida, Photocatalytic hydrogen production from water over a $LaMnO_3/CdS$ nanocomposite prepared by the reverse micelle method, J. Mater. Chem. 13 (2003) 1186-1191. https://doi.org/10.1039/B211812B

[83] P. Rai, Y.S. Kim, H.M. Song, M.K. Song, Y.T. Yu, The role of gold catalyst on the sensing behavior of ZnO nanorods for CO and $NO_2$ gases, Sens. Actuators B Chem. 165 (2012) 133-142. https://doi.org/10.1016/j.snb.2012.02.030

[84] M.F. Al-Kuhaili, S.M.A. Durrani, I.A. Bakhtiari, Carbon monoxide gas-sensing properties of $CeO_2$–ZnO thin films, Appl. Surf. Sci. 255 (2008) 3033-3039. https://doi.org/10.1016/j.apsusc.2008.08.058

[85] S. Arunkumar, T. Hou, Y.B. Kim, B. Choi, S.H. Park, S. Jung, D.W. Lee, Au Decorated ZnO hierarchical architectures: Facile synthesis, tunable morphology and enhanced CO detection at room temperature, Sens. Actuators B Chem. 243 (2017) 990-1001. https://doi.org/10.1016/j.snb.2016.11.152

[86] M.H. Huang, Y. Wu, H. Feick, N. Tran, E. Weber, P. Yang, Catalytic growth of zinc oxide nanowires by vapor transport, Adv. Mater. 13 (2001) 113-116. https://doi.org/10.1002/1521-4095(200101)13:2<113::AID-ADMA113>3.0.CO;2-H

[87] L. Vayssieres, Growth of arrayed nanorods and nanowires of ZnO from aqueous solutions, Adv. Mater. 15 (2003) 464-466. https://doi.org/10.1002/adma.200390108

[88] T. Krishnakumar, R. Jayaprakash, N. Pinna, N. Donato, A. Bonavita, G. Micali, G. Neri, CO gas sensing of ZnO nanostructures synthesized by an assisted microwave wet chemical route, Sens. Actuators B Chem. 143 (2009) 198-204. https://doi.org/10.1016/j.snb.2009.09.039

[89] P. Rai, Y.T. Yu, Synthesis of floral assembly with single crystalline ZnO nanorods and its CO sensing property, Sens. Actuators B Chem. 161 (2012) 748-754. https://doi.org/10.1016/j.snb.2011.11.027

[90] J.H. Kim, A. Mirzaei, H.W. Kim, S.S. Kim, Extremely sensitive and selective sub-ppm CO detection by the synergistic effect of Au nanoparticles and core–shell nanowires, Sens. Actuators B Chem. 249 (2017) 177-188. https://doi.org/10.1016/j.snb.2017.04.090

[91] N.H. Ha, D.D. Thinh, N.T. Huong, N.H. Phuong, P.D. Thach, H.S. Hong, Fast

response of carbon monoxide gas sensors using a highly porous network of ZnO nanoparticles decorated on 3D reduced graphene oxide, Appl. Surf. Sci. 434 (2018) 1048-1054. https://doi.org/10.1016/j.apsusc.2017.11.047

[92] F. Özütok, I.K. Er, S. Acar, S. Demiri, Enhancing the CO gas sensing properties of ZnO thin films with the decoration of MWCNTs, J. Mater. Sci.: Mater. Electron. 30 (2019) 259-265. https://doi.org/10.1007/s10854-018-0288-2

[93] D. Zhang, J. Wu, Y. Cao, Cobalt-doped indium oxide/molybdenum disulfide ternary nanocomposite toward carbon monoxide gas sensing, J. Alloys Compd. 777 (2019) 443-453. https://doi.org/10.1016/j.jallcom.2018.10.365

[94] H. Fu, C. Hou, F. Gu, D. Han, Z. Wang, Facile preparation of rod-like $Au/In_2O_3$ nanocomposites exhibiting high response to CO at room temperature, Sens. Actuators B Chem. 243 (2017) 516-524. https://doi.org/10.1016/j.snb.2016.11.162

[95] U. Lampe, J. Gerblinger, H. Meixner, Carbon-monoxide sensors based on thin films of BaSnO3, Sens. Actuators B Chem. 25 (1995) 657-660. https://doi.org/10.1016/0925-4005(95)85145-3

[96] E. Comini, G. Faglia, G. Sberveglieri, C. Cantalini, M. Passacantando, S. Santucci, Y. Li, W. Wlodarski, W. Qu, Carbon monoxide response of molybdenum oxide thin films deposited by different techniques, Sens. Actuators B Chem. 68 (2000) 168-174. https://doi.org/10.1016/S0925-4005(00)00484-6

[97] S. Vetter, S. Haffer, T. Wagner, M. Tiemann, Nanostructured $Co_3O_4$ as a CO gas sensor: Temperature-dependent behavior, Sens. Actuators B Chem. 206 (2015) 133-138. https://doi.org/10.1016/j.snb.2014.09.025

[98] R.J. Wu, J.G. Wu, T.K. Tsai, C.T. Yeh, Use of cobalt oxide CoOOH in a carbon monoxide sensor operating at low temperatures, Sens. Actuators B Chem. 120 (2006) 104-109. https://doi.org/10.1016/j.snb.2006.01.053

[99] A.K. Basu, P.S. Chauhan, M. Awasthi, S. Bhattacharya, $\alpha$-$Fe_2O_3$ loaded rGO nanosheets based fast response/recovery CO gas sensor at room temperature, Appl. Surf. Sci. 465 (2019) 56-66. https://doi.org/10.1016/j.apsusc.2018.09.123

[100] R. Kumar, M. Jaiswal, O. Singh, A. Gupta, M.S. Ansari, J. Mittal, Selective and reversible sensing of low concentration of carbon monoxide gas using Nb-doped OMS-2 nanofibers at room temperature, IEEE Sens. J. 19(2019) 7201-7206. https://doi.org/10.1109/JSEN.2019.2916485

[101] R. Kumar, N. Kushwaha, J. Mittal, Ammonia gas sensing activity of Sn nanoparticles film, Sensor Lett. 14) (2016) 300-303.

https://doi.org/10.1016/j.snb.2016.12.111

[102] J.C. Obirai, G. Hunter, P.K. Dutta, Multi-walled carbon nanotubes as high temperature carbon monoxide sensors, Sens. Actuators B Chem. 134 (2008) 640-646. https://doi.org/10.1016/j.snb.2008.06.005

[103] T.W. Ebbesen, Carbon nanotubes: preparation and properties, CRC press,Boca Raton,1996

[104] A.J. Page, F. Ding, S. Irle, K. Morokuma, Insights into carbon nanotube and graphene formation mechanisms from molecular simulations: a review, Rep. Prog. Phys. 78(3) (2015) 036501. https://doi.org/10.1088/0034-4885/78/3/036501

[105] M.S. Dresselhaus, G. Dresselhaus, P.C. Eklund, A.M. Rao, Carbon nanotubes, in: W. Andreoni (Ed.), The physics of fullerene-based and fullerene-related materials, Springer Netherlands, Dordrecht, 2000, pp. 331-379

[106] R.H. Baughman, A.A. Zakhidov, W.A. de Heer, Carbon nanotubes--the route toward applications, Science 297(5582) (2002) 787. https://doi.org/10.1126/science.1060928

[107] P. Kim, L. Shi, A. Majumdar, P.L. McEuen, Thermal transport measurements of individual multiwalled nanotubes,Phys. Rev. Lett. 87 (2001) 215502. https://doi.org/10.1103/PhysRevLett.87.215502

[108] M. Kociak, A.Y. Kasumov, S. Guéron, B. Reulet, I. Khodos, Y.B. Gorbatov, V. Volkov, L. Vaccarini, H. Bouchiat, Superconductivity in ropes of single-walled carbon nanotubes, Phys. Rev. Lett. 86 (2001) 2416-2419. https://doi.org/10.1016/S0921-4526(02)02148-8

[109] H.C. Su, N.V. Myung, Synthesis of platinum and tin oxide co-functionalized single-walled carbon nanotubes ($Pt/SnO_2/SWNTs$) and their sensing properties toward carbon monoxide, Electroanalysis 31 (2019) 437-447. https://doi.org/10.1002/elan.201800563

[110] S.C.K. Misra, P. Mathur, B.K. Srivastava, Vacuum-deposited nanocrystalline polyaniline thin film sensors for detection of carbon monoxide, Sens. Actuator A Phys. 114 (2004) 30-35. https://doi.org/10.1016/j.sna.2004.02.026

[111] W.R. Salaneck, I. Lundström, B.G. Rånby, Conjugated polymers and related materials: The interconnection of chemical and electronic structure: proceedings of the eighty-first Nobel Symposium, Oxford University Press, USA, 1993

[112] S. Radhakrishnan, S. Paul, Conducting polypyrrole modified with ferrocene for applications in carbon monoxide sensors, Sens. Actuators B Chem. 125 (2007) 60-

65. https://doi.org/10.1016/j.snb.2007.01.038

[113] S. Paul, F. Amalraj, S. Radhakrishnan, CO sensor based on polypyrrole functionalized with iron porphyrin, Synth. Met. 159 (2009) 1019-1023. https://doi.org/10.1016/j.synthmet.2009.01.018

[114] A. Gulino, T. Gupta, M. Altman, S. Lo Schiavo, P.G. Mineo, I.L. Fragalà, G. Evmenenko, P. Dutta, M.E. van der Boom, Selective monitoring of parts per million levels of CO by covalently immobilized metal complexes on glass, Chem. Commun. (2008) 2900-2902. https://doi.org/10.1039/B802670J

[115] M. Itou, Y. Araki, O. Ito, H. Kido, Carbon monoxide ligand-exchange reaction of triruthenium cluster complexes induced by photosensitized electron transfer: a new type of photoactive co color sensor, Inorg. Chem. 45 (2006) 6114-6116. https://doi.org/10.1021/ic060751n

[116] J. Esteban, J.V. Ros-Lis, R. Martínez-Máñez, M.D. Marcos, M. Moragues, J. Soto, F. Sancenón, Sensitive and selective chromogenic sensing of carbon monoxide by using binuclear rhodium complexes, Angew. Chem. 49 (2010) 4934-4937. https://doi.org/10.1002/anie.201001344

[117] A.R. Chakravarty, F.A. Cotton, D.A. Tocher, J.H. Tocher, Structural and electrochemical characterization of the novel ortho-metalated dirhodium(II) compounds $Rh_2(O_2CMe)_2[Ph_2P(C_6H_4)]_2$.cntdot.2L, Organometallics 4 (1985) 8-13. https://doi.org/10.1021/om00120a003

[118] J. Courbat, M. Pascu, D. Gutmacher, D. Briand, J. Wöllenstein, U. Hoefer, K. Severin, N.F. de Rooij, A colorimetric CO sensor for fire detection, Procedia Eng. 25 (2011) 1329-1332. https://doi.org/10.1016/j.proeng.2011.12.328

[119] C. Lin, X. Xian, X. Qin, D. Wang, F. Tsow, E. Forzani, N. Tao, High performance colorimetric carbon monoxide sensor for continuous personal exposure monitoring, ACS Sens. 3 (2018) 327-333. https://doi.org/10.1021/acssensors.7b00722

[120] A. Toscani, C. Marín-Hernández, M.E. Moragues, F. Sancenón, P. Dingwall, N.J. Brown, R. Martínez-Máñez, A.J.P. White, J.D.E.T. Wilton-Ely, Ruthenium(II) and osmium(ii) vinyl complexes as highly sensitive and selective chromogenic and fluorogenic probes for the sensing of carbon monoxide in air, Chem.: Eur. J. 21 (2015) 14529-14538. https://doi.org/10.1002/chem.201501843

[121] W. Feng, J. Hong, G. Feng, Colorimetric and ratiometric fluorescent detection of carbon monoxide in air, aqueous solution, and living cells by a naphthalimide-

based probe, Sens. Actuators B Chem. 251 (2017) 389-395.
https://doi.org/10.1016/j.snb.2017.05.099

[122] T.V. Dinh, I.Y. Choi, Y.S. Son, J.C. Kim, A review on non-dispersive infrared gas sensors: Improvement of sensor detection limit and interference correction, Sens. Actuators B Chem. 231 (2016) 529-538. https://doi.org/10.1016/j.snb.2016.03.040

[123] R. Diharja, M. Rivai, T. Mujiono, H. Pirngadi, Carbon monoxide sensor based on non-dispersive infrared principle, J. Phys. Conf. 1201 (2019) 012012. https://doi.org/10.1088/1742-6596/1201/1/012012

[124] C. Özbek, CO gas sensor applications of Fe doped calix[4]arene molecules, Izmir Institute of Technology, Izmir, Turkey, 2013

[125] J.W. Gardner, P.K. Guha, F. Udrea, J.A. Covington, CMOS interfacing for integrated gas sensors: A review, IEEE Sens. J. 10 (2010) 1833-1848. https://doi.org/10.1109/JSEN.2010.2046409

[126] G. Hanrahan, D.G. Patil, J. Wang, Electrochemical sensors for environmental monitoring: design, development and applications, J. Environ. Monit. 6 (2004) 657-664. https://doi.org/10.1039/b403975k

[127] H.D. Wiemhöfer, W. Göpel, Fundamentals and principles of potentiometric gas sensors based upon solid electrolytes, Sens. Actuators B Chem. 4 (1991) 365-372. https://doi.org/10.1016/0925-4005(91)80137-9

[128] P. Moseley, Solid state gas sensors, Meas. Sci. Technol. 8 (1997) 223. https://doi.org/10.1088/0957-0233/8/3/003

[129] J. Liu, W. Weppner, Beta″-alumina solid electrolytes for solid state electrochemical $CO_2$ gas sensors, Solid State Commun. 76 (1990) 311-313. https://doi.org/10.1016/0038-1098(90)90844-2

[130] T. Ogata, S. Fujitsu, M. Miyayama, K. Koumoto, H. Yanagida, $CO_2$ gas sensor using$\beta$-$Al_2O_3$ and metal carbonate, J. Mater. Sci. Lett. 5 (1986) 285-286. https://doi.org/10.1007/BF01748079

[131] W. Kingery, J. Pappis, M. Doty, D. Hill, Oxygen ion mobility in cubic $Zr_{0.85}.Ca0.15O_{1.85}$, J. Am. Ceram. Soc. 42 (1959) 394. https://doi.org/10.1111/j.1151-2916.1959.tb13599.x

[132] H. Okamoto, H. Obayashi, T. Kudo, Carbon monoxide gas sensor made of stabilized zirconia, Solid State Ion. 1 (1980) 319-326. https://doi.org/10.1016/0167-2738(80)90012-0

[133] B. Yang, C. Wang, R. Xiao, H. Yu, J. Wang, J. Xu, H. Liu, F. Xia, J. Xiao, CO Response characteristics of $NiFe_2O_4$sensing material at elevated temperature, J. Electrochem. Soc. 166 (2019) B956-B960. https://doi.org/10.1149/2.0581912jes

[134] N. Miura, T. Raisen, G. Lu, N. Yamazoe, Zirconia-based potentiometric sensor using a pair of oxide electrodes for selective detection of carbon monoxide, J. Electrochem. Soc. 144 (1997) L198-L200. https://doi.org/10.1149/1.1837798

[135] J.Y. Park, S.J. Song, E.D. Wachsman, Highly sensitive/selective miniature potentiometric carbon monoxide gas sensors with titania-based sensing elements, J. Am. Ceram. Soc. 93 (2010) 1062-1068. https://doi.org/10.1111/j.1551-2916.2009.03500.x

[136] K. Mahendraprabhu, A. Selva Sharma, P. Elumalai, CO sensing performances of YSZ-based sensor attached with sol-gel derived ZnO nanospheres, Sens. Actuators B Chem. 283 (2019) 842-847. https://doi.org/10.1016/j.snb.2018.11.164

[137] S.A. Anggraini, V.V. Plashnitsa, P. Elumalai, M. Breedon, N. Miura, Stabilized zirconia-based planar sensor using coupled oxide(+Au) electrodes for highly selective CO detection, Sens. Actuators B Chem. 160 (2011) 1273-1281. https://doi.org/10.1016/j.snb.2011.09.062

[138] T. Hyodo, M. Takamori, T. Goto, T. Ueda, Y. Shimizu, Potentiometric CO sensors using anion-conducting polymer electrolyte: Effects of the kinds of noble metal-loaded metal oxides as sensing-electrode materials on CO-sensing properties, Sens. Actuators B Chem. 287 (2019) 42-52. https://doi.org/10.1016/j.snb.2019.02.036

[139] R. Knake, P. Jacquinot, A.W.E. Hodgson, P.C. Hauser, Amperometric sensing in the gas-phase, Anal. Chim. Acta 549 (2005) 1-9. https://doi.org/10.1016/j.aca.2005.06.007

[140] Z. Cao, W.J. Buttner, J.R. Stetter, The properties and applications of amperometric gas sensors, Electroanalysis 4 (1992) 253-266. https://doi.org/10.1002/elan.1140040302

[141] Z. Cao, W.J. Buttner, J.R. Stetter, The properties and applications of amperometric gas sensors, Electroanalysis 4 (1992) 253-266. https://doi.org/10.1002/elan.1140040302

[142] T. Otagawa, M. Madou, S. Wing, J. Rich-Alexander, S. Kusanagi, T. Fujioka, A. Yasuda, Planar microelectrochemical carbon monoxide sensors, Sens. Actuators B Chem. 1 (1990) 319-325. https://doi.org/10.1016/0925-4005(90)80223-M

[143] H. Yan, C.C. Liu, A solid polymer electrolyte-bases electrochemical carbon monoxide sensor, Sens. Actuators B Chem. 17 (1994) 165-168. https://doi.org/10.1016/0925-4005(94)87045-4

[144] P.D. Van der Wal, N.F. de Rooij, M. Koudelka-Hep, Extremely stable Nafion based carbon monoxide sensor, Sens. Actuators B Chem. 35 (1996) 119-123. https://doi.org/10.1016/S0925-4005(97)80040-8

[145] P. Santhosh, K.M. Manesh, A. Gopalan, K.P. Lee, Novel amperometric carbon monoxide sensor based on multi-wall carbon nanotubes grafted with polydiphenylamine—Fabrication and performance, Sens. Actuators B Chem. 125 (2007) 92-99. https://doi.org/10.1016/j.snb.2007.01.044

[146] C.H. Chou, J.L. Chang, J.M. Zen, Homogeneous platinum-deposited screen-printed edge band ultramicroelectrodes for amperometric sensing of carbon monoxide, Electroanalysis 21 (2009) 206-209. https://doi.org/10.1002/elan.200804376

[147] A.F. Ismail, K.C. Khulbe, T. Matsuura, RO membrane characterization, in: A.F. Ismail, K.C. Khulbe, T. Matsuura (Eds.), Reverse osmosis, Elsevier, Amsterdam, Netherlands, 2019, pp. 57-90

[148] M. Forster, Optoacoustic gas sensor, Google Patents, 1999, US6006585A

[149] N. Petra, J. Zweck, A.A. Kosterev, S.E. Minkoff, D. Thomazy, Theoretical analysis of a quartz-enhanced photoacoustic spectroscopy sensor, Appl. Phys. B 94 (2009) 673-680. https://doi.org/10.1007/s00340-009-3379-1

[150] Y. Ma, Y. Tong, Y. He, X. Jin, F.K. Tittel, Compact and sensitive mid-infrared all-fiber quartz-enhanced photoacoustic spectroscopy sensor for carbon monoxide detection, Opt. Express 27 (2019) 9302-9312. https://doi.org/10.1364/OE.27.009302

[151] L. Tao, K. Sun, M.A. Khan, D.J. Miller, M.A. Zondlo, Compact and portable open-path sensor for simultaneous measurements of atmospheric $N_2O$ and CO using a quantum cascade laser, Opt. Express 20 (2012) 28106-28118. https://doi.org/10.1364/OE.20.028106

[152] Y. Ma, G. Yu, J. Zhang, X. Yu, R. Sun, Sensitive detection of carbon monoxide based on a QEPAS sensor with a 2.3 μm fiber-coupled antimonide diode laser, J. Opt. 17 (2015) 055401. https://doi.org/10.1088/2040-8978/17/5/055401

[153] X. Yin, L. Dong, H. Zheng, X. Liu, H. Wu, Y. Yang, W. Ma, L. Zhang, W. Yin, L. Xiao, Impact of humidity on quartz-enhanced photoacoustic spectroscopy based CO detection using a near-IR telecommunication diode laser, Sensors 16 (2016)

162. https://doi.org/10.3390/s16020162

[154] Y. He, Y. Ma, Y. Tong, X. Yu, F.K. Tittel, A portable gas sensor for sensitive CO detection based on quartz-enhanced photoacoustic spectroscopy, Opt. Laser Technol. 115 (2019) 129-133. https://doi.org/10.1016/j.optlastec.2019.02.030

[155] S. Li, L. Dong, H. Wu, A. Sampaolo, P. Patimisco, V. Spagnolo, F.K. Tittel, Ppb-level quartz-enhanced photoacoustic detection of carbon monoxide exploiting a surface grooved tuning fork, Anal. Chem. 91 (2019) 5834-5840. https://doi.org/10.1021/acs.analchem.9b00182

[156] P. Mehrotra, Biosensors and their applications - A review, J. Oral. Biol. Craniofac. Res. 6 (2016) 153-159. https://doi.org/10.1016/j.jobcr.2015.12.002

[157] Y. Cui, B. Lai, X. Tang, Microbial fuel cell-based biosensors, Biosensors 9 (2019) 92. https://doi.org/10.3390/bios9030092

[158] C.H. Yeh, Y.H. Chang, H.P. Lin, T.C. Chang, Y.C. Lin, A newly developed optical biochip for bacteria detection based on DNA hybridization, Sens. Actuators B Chem. 161 (2012) 1168-1175. https://doi.org/10.1016/j.snb.2011.10.016

[159] S.N. Sawant, Development of biosensors from biopolymer composites, in: K.K. Sadasivuni, D. Ponnamma, J. Kim, J.J. Cabibihan, M.A. AlMaadeed (Eds.), Biopolymer composites in electronics, Elsevier, Amsterdam, Netherlands, 2017, pp. 353-383

[160] O. Soldatkin, I. Kucherenko, V. Pyeshkova, A. Kukla, N. Jaffrezic-Renault, A. El'Skaya, S. Dzyadevych, A. Soldatkin, Novel conductometric biosensor based on three-enzyme system for selective determination of heavy metal ions, Bioelectrochemistry 83 (2012) 25-30. https://doi.org/10.1016/j.bioelechem.2011.08.001

[161] P.J. Conroy, S. Hearty, P. Leonard, R.J. O'Kennedy, Antibody production, design and use for biosensor-based applications, Semin. Cell Dev. Biol. 20 (2009) 10-26. https://doi.org/10.1016/j.semcdb.2009.01.010

[162] W. Yi-Xian, Y. Zun-Zhong, S. Cheng-Yan, Y. Yi-Bin, Application of aptamer based biosensors for detection of pathogenic microorganisms, Chin. J. Anal. Chem. 40 (2012) 634-642. https://doi.org/10.1016/S1872-2040(11)60542-2

[163] S. Melamed, T. Elad, S. Belkin, Microbial sensor cell arrays, Curr. Opin. Biotechnol. 23 (2012) 2-8. https://doi.org/10.1016/j.copbio.2011.11.024

[164] C.L. Nobles, J.R. Clark, S.I. Green, A.W. Maresso, A dual component heme biosensor that integrates heme transport and synthesis in bacteria, J.

Microbiol.Methods 118 (2015) 7-17. Https://doi.org/10.1016/j.mimet.2015.07.011

[165] M.K. Chan, Heme protein biosensors, J. Porphyr. Phthalocyanines 04 (2000) 358-361. https://doi.org/10.1002/(SICI)1099-1409(200006/07)4:4<358::AID-JPP243>3.0.CO;2-A

[166] L. Gorton, A. Lindgren, T. Larsson, F. Munteanu, T. Ruzgas, I. Gazaryan, Direct electron transfer between heme-containing enzymes and electrodes as basis for third generation biosensors, Anal. Chim. Acta 400 (1999) 91-108. https://doi.org/10.1016/S0003-2670(99)00610-8

[167] D. Shelver, R.L. Kerby, Y. He, G.P. Roberts, CooA, a CO-sensing transcription factor from Rhodospirillum rubrum, is a CO-binding heme protein, Proc. Natl. Acad. Sci. USA 94 (1997) 11216-11220. https://doi.org/10.1073/pnas.94.21.11216

[168] B.R. Weaver, R.M. DeVries, A.J. Gunter, R.W. Clark, Unexpected pH-dependent DNA Binding of CooA, a CO-Sensing Transcription Factor from Rhodospirillum rubrum, FASEB J. 30 (2016) 835.7-835.7. https://doi.org/10.1096/fasebj.30.1_supplement.835.7

# About the authors

## *Gurleen Kaur Gulati*

The author received a BSc (Hons.) in Chemistry from St. Stephen's College, University of Delhi (India) in 2013, and an MSc in Chemistry from the University of Delhi in 2015. She is working as a research fellow under the supervision of Dr. Satish Kumar at St. Stephen's College. The research involved the synthesis of macrocyclic receptors for the detection of toxic ions.

## *Loveleen Kaur Gulati*

The author received a BSc (Hons.) in Chemistry from St. Stephen's College, University of Delhi (India) in 2013, and an MSc in Chemistry from the University of Delhi in 2015. She is working as a research fellow under the supervision of Dr. Satish Kumar at St. Stephen's College. The research aimed at the development of photochromic receptors for toxic analyte sensing.

## *Satish Kumar*

The author is an assistant professor in the Department of Chemistry at St. Stephen's College, University of Delhi. During the last 18 years, Dr. Satish Kumar has worked in interdisciplinary areas covering theoretical chemistry, nanochemistry, and application of principles of molecular recognition to design molecular receptors. Dr. Satish Kumar has published several research papers related to the development of receptors for neutral molecules, anions, and cations. He has 16 years of teaching experience.